Research Notes in Neural Computing

U. Grenander Y. Chow D.M. Keenan

Hands

A Pattern Theoretic Study of Biological Shapes

With 92 Illustrations, 43 in Full Color

Springer-Verlag
New York Berlin Heidelberg London
Paris Tokyo Hong Kong Barcelona

U. Grenander
Division of
Applied Mathematics
Brown University
Providence, RI 02912
USA

Y. Chow
Academia Sinica
Institute of Mathematics
Taipei, Taiwan R.O.C.

D.M. Keenan
Department of Mathematics
Univeristy of Virginia
Charlottesville, VA 22903
USA

Managing Editor
Bart Kosko
Department of Electrical Engineering
Signal and Image Processing Institute
University of Southern California
Los Angeles, CA 90089-0782
USA

Printed on acid-free paper.

Photocomposed copy prepared from the authors' *AMS*TEX file.
Printed and bound by R.R. Donnelley & Sons, Harrisonburg, VA.
Printed in the United States of America.

9 8 7 6 5 4 3 2 1

ISBN 0-387-97386-9 Springer-Verlag New York Berlin Heidelberg
ISBN 3-540-97386-9 Springer-Verlag Berlin Heidelberg New York

Preface

In this book a global shape model is developed and applied to the analysis of real pictures acquired with a visible light camera under varying conditions of optical degradation. Computation feasibility of the algorithms derived from this model is achieved by analytical means.

The model is based on general pattern theoretic ideas specialized to pattern inference for noisy images.

The aim is to develop methods for image understanding based on *structured* restoration. We also want to find the limits of applicability of the algorithms. This is done by making the optical degradations more and more severe until the algorithms no longer succeed in their task.

This computer experiment in pattern theory is one of several. The others, LEAVES, X-RAYS, and RANGE are described elsewhere.

Acknowledgments

Supported by the National Science Foundation Contract Nos. DMS 86020-38, DMS 8813699, the Army Research Office Contract Nos. DAAG-29-83-K-0116, DAAL-03-86-K-0171, and the Office of Naval Research Contract Nos. N00014-87-K-0367, N00014-88-K-0289, and N00014-90-J-1007.

Yunshyong Chow was on leave from the Institute of Mathematics, Academia Sinica, Taipei, Taiwan, R.O.C.

Daniel MacRae Keenan is now with the Department of Mathematics, University of Virginia, Charlottesville.

AMS-TEX is the TEX macro system of the American Mathematical Society.

We received much help from Dr. Alan Knoerr and Daniel F. Potter both with respect to the programming aspect and on the technique of image acquisition.

Ezoura Fonseca typed the manuscript using TEX system of the American Mathematical Society and did so quickly and with great expertise. We would also like to thank Felisa Vázquez-Abad for her help.

Contents

1 Dealing with Highly Variable Patterns

Is it possible to mechanize human intuitive understanding of biological pictures that typically exhibit a lot of variability but also possess characteristic structure?

To clarify what we mean by 'understanding' let us consider the following tasks in pattern analysis.

(i) Given a pure image I from some ensemble $\mathcal{I}$, observed by some technology that produces a noisy version of it, say a deformed image $I^{\mathcal{D}}$, can we construct an algorithm that restores I from $I^{\mathcal{D}}$ so well that we judge the restored picture I^* to be a reasonably faithful copy of I?

(ii) Same, but given a subset $\mathcal{I}_{abnorm} \subset \mathcal{I}$ can we design a test that discovers when $I \in \mathcal{I}_{abnorm}$ with accuracy comparable to that of human judgment and that locates the *pathology* in the shape?

(iii) Same, but with an objectively defined *numerical characteristic* for the elements in the pure image ensemble $\mathcal{I}$, can we estimate them as well from $I^{\mathcal{D}}$ as a human can do?

In this paper we shall report the results of an experiment dealing with questions (i) and (ii). A parallel experiment in Knoerr (1988) has been carried out to study (iii) for leaf shapes. A third experiment to investigate (i) for X-ray pictures has been completed and will be reported elsewhere.

To get some feeling for what are the main difficulties consider the digital hand picture in Plate 1(a) (we use the convention that a,b,c,d refer to the pictures in the upper left, upper right, lower left, and lower right respectively). It shows a 128×120 picture (CDA6) which leads after thresholding to Plate 1(c). The histogram of the grey level values in the original picture Plate 1(a) is displayed in Plate 2 along with a profile along a line. Plate 1(b) is a relatively noiseless picture with the same hand and hand position as Plate 1(a). Thresholding and boundary extraction of Plate 1(b) lead to Plate 1(d). In Plates 2 and 3 the extracted boundary has been superimposed to enhance visual identification.

While Plate 1(a) seems to be fairly clear it is seen, when examining

Plate 2, that the visual noise hides the local features of the boundary. The histogram should ideally have two well separated peaks, one at a low intensity for the area inside the hand, and another at a higher intensity outside the hand. The two peaks can perhaps be seen but would be difficult to separate by an algorithm. It is also clear that the line profile gives only weak evidence about where the fingers are.

An even noisier picture is shown in Plate 39(a); when displayed on a screen most observers barely see the hand shape, some not at all.

In such situations, can we design an objective method for restoring the true shape and its structure with a performance that compares well with that of human visual processing?

§1.1. The pattern recognition literature is full of techniques developed for image restoration. One could apply some of them, for example the following sequence, to our task(s),

a) remove some of the noise by filtering the picture, say by 3×3 median filter,

b) apply a filling/thinning procedure to further reduce the noise by filling holes and removing pixels that are more or less isolated,

c) threshold the picture,

d) use an edge detector to compute the boundary,

e) close the boundary, if needed,by closing small gaps,

f) compute the connected components of the boundary, keep one, probably the largest one, as the estimate of the hand contour.

This procedure, or one of its many possible variations, is not unreasonable. It is well known, however, that it will not work well unless precautions are taken during the capture of the image. For example, the thresholding will not yield acceptable results if lighting is non-uniform, or the boundaries obtained will be too irregular unless smoothed, and cannot be guaranteed to be hand-like.

A procedure of this type is likely to produce hand-like restorations $I*$, only for a large S/N ratios as in Plate 1(b), where in Plate 3 the profile across the fingers has clear high-low regions and the histogram has two well separated peaks.

There is no guarantee, of course, that $I*$ will have a thumb, four other fingers, a palm etc., since the algorithm does not know that human hands have those features. It will therefore not be powerful enough to carry out tasks that refer to *biologically motivated concepts*: tip of little finger, length of thumb, etc.

If we insist that $I*$ should have such hand-like features we must build some *prior knowledge of human hands*, or whatever the biology is, into the algorithm. We could construct a template and fit it by least squares or more sophisticated techniques to $I^{\mathcal{D}}$; this is one of the oldest methods of pattern recognition and consists just of a matched filter. We would then have to estimate three parameters: two for location and one for orientation. Or, we may add another degree of freedom for scale, so that we have four degrees of freedom.

We could then be sure that $I*$ will have a "real" hand shape, but at the price of losing information about the smaller features, lengths of digits, angles between them, and so on.

These two approaches represent extremes in that the first one is general purpose and is intended for arbitrary shapes, while the second one is tailor made for a special picture ensemble.

We propose instead that the *restoration should be based on a mathematical model that incorporates biological knowledge about the shape including its variability.* The latter should be emphasized since in some applications the interest is in the anomalies, deviations from a norm, rather than in the typical shape.

§1.2. To do this we must allow for the fact that we are dealing with high dimensional hypothesis spaces. In other situations dealing with man made objects this need not be the case. If the shapes belong to a solid and rigid body we need only consider variability due to actions of the Euclidean group, in other cases perhaps some other low-dimensional Lie group will take its place. Discretizing we may be able to list the shapes in a table and base the algorithm on table look up modulo the group in question.

For biological shapes this is not acceptable. Instead we shall use a pattern theoretic approach, constructing *powerful models of random shape that express prior knowledge about the variability and global properties of the images.* For a general discussion of such models the reader is referred to Grenander (1981), Chapter 5, and specialized to the case of random shape to Grenander-Keenan (1986).

The general pattern theoretic model defines a configuration as an element of a regular structure $\mathcal{C}$

$$c = \sigma(g_0, g_1, \dots g_i \dots g_{n-1}) \in \mathcal{C} \tag{1.1}$$

where σ is the connector graph (from some family Σ of graphs, the connection type) with sites $i = 0, 1, \dots n-1$ and the g's are generators from some generator space G. The generators carry bond values, $\beta_j(g)$ from some bond value space B; $j = 0, 1, \dots \omega(g) - 1$; where $\omega(g)$ is the arity of g, and we shall only consider regular configurations satisfying *the first structure formula* with a logical conjunction

$$\wedge_\sigma \rho[\beta_{j_1}(g_{i_1}), \beta_{j_2}(g_{i_2})] = TRUE \tag{1.2}$$

with a Boolean bond value function

$$\rho : B \times B \to \{TRUE, FALSE\}$$

and where the conjunction in (1.2) is over all bond pairs $(i_1, j_1) - (i_2, j_2)$ that appear in the connector graph σ.

In addition to the *local regularity* expressed through the bond value function ρ we may have a *global regularity* condition γ and we shall write $\mathcal{R} =< \Sigma, \rho, \gamma >$ for the full regularity. The set C of all configurations c for which $\mathcal{R}$ holds constitute the (regular) *configuration space* $\mathcal{C}$.

Let us make this concrete by the following specialization that will be adhered to in the present study:

$$(1.3) \qquad \begin{cases} \Sigma = \text{all cyclic graphs} \\ g = \text{directed } C^2 - \textit{arc in } \boldsymbol{R}^2 \\ \beta_0(g) = \text{initial point of } \textit{arc } g \\ \beta_1(g) = \text{end point of } \textit{arc } g \\ \omega(g) \equiv 2 \\ \rho(\beta', \beta'') = \text{TRUE iff } \beta' = \beta'' \end{cases}$$

This means that we join n arcs together to form a closed curve in $\boldsymbol{R}^2$. When needed we shall impose the global constraint

$$(1.4) \qquad \gamma(c) = \text{TRUE iff } \ c \text{ is not self-intersecting.}$$

The configuration space can then be viewed as a family of simple arc polygons in the plane.

Any $c \in \mathcal{C}(\mathcal{R})$, describing a Jordan curve in the plane, defines a set $I =$ the interior of the curve which we shall speak of as a *pure image.* The set of such I's forms the image algebra $\mathcal{I}$ and we have a mapping:

$$(1.5) \qquad R : \mathcal{C}(\mathcal{R}) \longrightarrow \mathcal{I}$$

which does not always have an inverse, a fact that causes a good deal of trouble in some applications.

§1.3. The next step is to introduce a probability measure, a prior, on $\mathcal{C}(\mathcal{R})$ which will play the role of a parameter space in the following Bayesian approach. To do this we shall start from *acceptor functions* A_i defined on $S \times S$ and taking values in $\boldsymbol{R}_+$, where S is a *group of similarity transformations* $s : G \rightarrow G$. The similarity group S will be chosen as a low-dimensional Lie group which in the present case will be specialized to one of the following:

(i) the group of uniform scale change US(2) in the plane: $s(x, y) = (cx, cy)$; $c \neq 0$;

(ii) the general linear group GL(2) in $\boldsymbol{R}^2$

$$s(x, y) = \begin{pmatrix} s_{00} & s_{01} \\ s_{10} & s_{11} \end{pmatrix} \begin{pmatrix} s \\ y \end{pmatrix}; \ det \begin{pmatrix} s_{00} & s_{01} \\ s_{10} & s_{11} \end{pmatrix} \neq 0$$

(iii) the product US(2) ×0(2), 0(2) = orthogonal group in $\boldsymbol{R}^2$; the product group can be parametrized as

$$s(x,y) = \begin{pmatrix} u & v \\ -v & u \end{pmatrix} \begin{pmatrix} x \\ y \end{pmatrix}; (u,v) \neq 0.$$

If $g \in G$ is an arc in the plane sg shall mean the set that results when letting s operate on each point of g. We must of course assume that G is invariant under $S: sg \in G, \forall s \in S, g \in G$.

Let m be a fixed measure on S and introduce the density with respect to m by *the second structure formula*

$$p(s_0, s_1, \ldots s_{n-1}) = \frac{1}{Z} \prod_{i=0}^{n-1} A_i(s_i, s_{i+1}) \tag{1.6}$$

where Z is a normalization constant to make $P(S^n) = 1$. In order that (1.6) be a legitimate density we shall assume that $A \in L_2(S \times S)$.

It will be convenient later on to write (1.6) in other forms, one of which is

$$A_i(s_i, s_{i+1}) = Q_i(s_i) A_i^0(s_i, s_{i+1}). \tag{1.7}$$

In (1.7) the meaning of A_i^0 is to express the stochastic coupling between s_i and s_{i+1} while $Q_i(s_i)$ is a weight function that makes different values in the similarity group S more or less likely.

Sometimes we shall write

$$A_i^0(s_i, s_{i+1}) = \overline{A}_i \left(\frac{s_i - s_{i+1}}{\epsilon} \right), \; \epsilon > 0, \tag{1.8}$$

where ϵ is a parameter expressing the strength of the coupling: small ϵ means strong coupling. For (1.8) to be meaningful we must assume that S can be made into a vector space, which is the case for the above choices (i), (ii), and (iii).

It is well known that the density in (1.6) is of Gibbs type and *defines a Markov process (in a cyclic sense)* $s_0, s_1, \ldots s_{n-1}$, see for example Grenander and Keenan (1986). We now restrict the measure by introducing a template

$$c_{temp} = \sigma(g_0^0, g_1^0, \ldots g_{n-1}^0) \in \mathcal{C}(\mathcal{R}) \tag{1.9}$$

and considering only those s-sequences for which

$$c = \sigma(s_0 g_0^0, s_1 g_1^0, \ldots s_{n-1} g_{n-1}^0) \in \mathcal{C}(\mathcal{R}) \tag{1.10}$$

For the special situations that will be encountered below, this conditioning of the probability measure is easily made, but it deserves mentioning that in a general context the conditioning should not be made by the usual (Kolmogorov) construction of conditional probabilities. Instead one should use the "old-fashioned" way of a limit argument made rigorous; this is discussed in Grenander (1981), Section 5.2.

This will result in a measure, say P, on the configuration space $\mathcal{C}(\mathcal{R})$ and the mapping R induces a probability measure on the *image algebra* $\mathcal{I}$ of random sets. P is going to be our prior measure and represents random shape with given global properties:

a) *the template* (or sometimes templates) c_{temp} *expresses the typical overall architecture of the shape,*

b) *the group-valued Markov process* $(s_i;\ i = 0, 1, \ldots n-1)$ *expresses random variation of the shape.*

It remains to describe how a pure image $I \in \mathcal{I}$ gives rise to an observed (deformed) image $I^{\mathcal{D}}$ in terms of a conditional probability measure

$$P(I^{\mathcal{D}} \in F|I);\ F \text{ Borel set in } \mathcal{I}^{\mathcal{D}} \tag{1.11}$$

where $\mathcal{I}^{\mathcal{D}}$ is the set of possible deformed images. For example, if we use an $LV \times LH$ digital camera with NL grey levels, where NL is large, it is natural to think of $\mathcal{I}^{\mathcal{D}}$ as embedded in $\boldsymbol{R}^{NV \times NL}$, as we shall do in the following.

The analytical form of (1.11) depends of course upon what technology is used to acquire the digital pictures. We shall postpone the discussion of this until Section 3.

Here we are dealing with *two sorts of randomness.* The global shape model describes the biological variability, both between individuals and for the possible hand shapes of a given individual. This randomness is intrinsic to the patterns we are dealing with and must play an important role in any serious study of variable shape. The other randomness is due to imperfections in the instrumentation and is a technological artifact; it is easier to study than the first one since it is more familiar.

The idea of a flexible template goes back to d'Arcy Thompson (1917). The formalization in terms of probabilistic deformations was introduced in Grenander (1970).

Different but somewhat related work can be found in Kendall's shape theory, Kendall (1977), and in Bookstein's biorthogonal grids, Bookstein (1978).

§1.4. Assuming the model to be completely specified, as it will be in Section 3.1, it is easy to see, at least in principle, how to apply it. For example, to use it for pattern synthesize we simulate the prior density on $\mathcal{C}(\mathcal{R})$ induced by the random group elements in (1.6) and (1.10). Or, to carry out image restoration, we simulate the posterior density

$$p(c|I^{\mathcal{D}}) \propto p(c)p(I^{\mathcal{D}}|c) \tag{1.12}$$

using (1.6), (1.10), and (1.11). If we produce an i.i.d. sample $(c_1, c_2, \dots c_N)$ from (1.12) we can compute an average

$$c^* = \frac{1}{N} \sum_{\nu=1}^{N} c_\nu \tag{1.13}$$

and use $I^* = Rc^*$ as an estimate of the true set I that was to be restored. Note that the configuration space here can be embedded in a vector space. This procedure also gives us information about the accuracy that should be attributed to our restoration c^* from the sample $c_1, c_2, \dots c_N$.

Alternatively we can simulate a density proportional to $[p(c|I^{\mathcal{D}})]^{1/T}$, where T, the temperature, is allowed to decrease slowly to zero according to some appropriate schedule; this is the simulated annealing approach that has been applied successfully to image restoration in Geman-Geman (1984) and Gidas (1985), and gives us the (asymptotically) MAP estimator.

To realize the above we shall use stochastic relaxation. Since our generator space is now a continuum it is necessary to justify it rigorously in this new context; it will be done in Section 2.

To actually implement such Monte Carlo simulations requires careful planning to avoid unacceptable CPU-time requirement. At first glance it may appear that the continuum-based model used here should require much less computing than the lattice-based ones earlier in for example Besag (1986), Geman-Geman (1984), Grenander (1984), Ripley (1986): after all, the number of sites, n, which in the lattice models is of the order $10^4 - 10^6$, need only be a few hundred, or even less, for the shape model. This is indeed the case, but this advantage is offset, at least partly, by the increased computational effort needed for updating each site.

We shall see in Sections 2 and 3 *how these computational obstacles can be overcome to a considerable degree by analytical means.*

2
Pattern Synthesis

The prior induced by the Gibbs density (1.6) cannot be simulated directly for a general acceptor function and for n-values of practical interest. We shall therefore apply stochastic relaxation that has earlier been used extensively for lattice based models; see Geman-Geman (1984), Grenander (1983).

§2.1. Consider a sequence $SW_1, SW_2, \dots$ of subsets in $\{0, 1, \dots n-1\}$, *sweep areas*, that may overlap, and such that their union is equal to the set $\{0, 1, \dots n-1\}$ of all sites in the connector graph.

Also introduce a *sweep strategy* as a sequence $SW_{j_1}, SW_{j_2}, \dots$ of sweep areas, together covering all sites, and such that all sites are visited infinitely often.

Stochastic relaxation is organized as follows. Initialize by picking a starting configuration $c(0)$. In principle this choice is arbitrary, but the speed of convergence is influenced by it.

At iteration no. t, we have arrived at a configuration $c(t)$, and now modify it by calculating the conditional probability density using (1.6) and the imposed regularity constraints as

$$p[c_{|SW_{j(t)}} | \bar{c}(t)] \tag{2.1}$$

where $c_{|SW_{j(t)}}$ is the subconfiguration of c restricted to $SW_{j(t)}$ and $\bar{c}(t)$ is the subset of $c(t)$ where

$$g_i = g_i(t) \text{ for } i \notin SW_{j(t)}$$

Simulate the density (2.1) independently of what has happened before and place the resulting subconfiguration at the sites $i \in SW_{j(t)}$.

Iterate this updating procedure for $t = 1, 2, \dots$. In the next sections we shall make the claim, under specified conditions, that the probability measures for the randomized configurations $c(t)$ tend weakly to the measure induced by (1.6).

In all cases to be considered in the hand shape study the generator space G will be embedded in some finite dimensional Euclidean space so that $\mathcal{C}(\mathcal{R})$ will also be a subset of some $\boldsymbol{R}^d$ where the dimension d is large but finite. Then the random sequence $c(0), c(1), c(2), \ldots$ forms a Markov chain in time with a continuum as its state space.

Among the many possible sweep strategies the following two deserve special attention. For the first one, let sweep areas consist of single sites $SW = \{j\};\ j = 0, 1, \ldots n-1;$ and visit them in cyclic order $j(t) \equiv t(mod\ n)$. It is then natural to choose n updates as our time unit, a full sweep.

The second strategy is to pick, at each update, one of the n sites at random with equal probabilities. If, at time $t+1$, one happens to get the same site as at time t one should discard the site and try again. Indeed, updating the same site twice in a row has the same effect (probabilistically) as doing it once.

For both of these sweep strategies $c(t)$ forms a stationary Markov chain in time. Note that the prior measure is an equilibrium distribution since if the random configuration $c(t)$ is distributed according to the prior then $c(t+1)$ has the same distribution.

If we can prove that the chain is ergodic then the stochastic relaxation procedure converges weakly to the equilibrium measure which coincides with the prior. To prove ergodicity, when we have an equilibrium measure as above, we first need to establish irreducibility (see Feller (1966), section VIII.7), any two states are *pathwise connected* in the probabilistic sense. The authors at first believed that connectivity would be obvious for all three similarity groups considered in this study, but this was not so, and the proofs to be given in the next section and Appendix D were arrived at after several unsuccessful attempts. As a consequence of our connectivity results (given below and in Appendix D) we have that the chain, $\{c(t)\}_{t=0}^{\infty}$, is ν-irreducible with ν being Lebesgue measure (see Revuz (1984), p. 87). From above we know that the chain created without imposing the closure and nonselfintersecting conditions has a bounded invariant measure, μ. Since it is ν-irreducible and aperiodic, it follows from Theorem 2.3, p. 88 and Exercise 2.15, p. 99, Revuz (1984), that this chain is recurrent and thus by Prop. 2.5, p. 198, Revuz (1984), it is ergodic. The conditional probability measure constructed from μ is a bounded invariant measure and thus $\{c(t)\}_{t=0}^{\infty}$ is ergodic.

§2.2. We shall consider three choices of the similarity group S, the first of which is uniform scale change, $S = US(2)$. Remember that we are dealing with closed simple curves, here polygons.

In Figure 1(a) we display a template with $n = 12$. It is clear that the polygon in (b) belongs to the same shape class under $S = US(2)$. Indeed, in (a) we can first lower the values of s_1, s_3, then lower the values of s_7, s_9, then increase s_{10}, s_4 while lowering

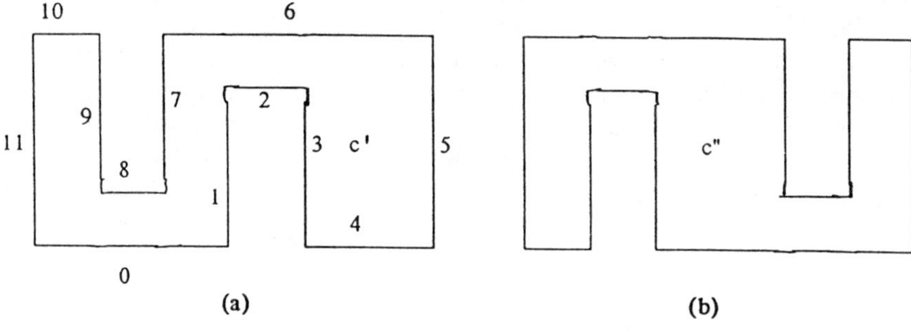

FIGURE 1

s_0, s_6, and finally change s_1, s_3, s_7, s_9 back to the original values - this gives us (b) obtained as a continuous deformation of (a). Hence (a) and (b) are connected by a continuous path contained in the shape class.

However, if the acceptor functions are such that $A(s', s'') = 0$ except when s' and s'' are close to the identity then the two configurations c' and c'' are not connected in terms of the support of the prior measure.

On the other hand if $A(s', s'') > 0$ for all positive values of the scale factors s' and s'' it seems plausible that the configuration space is connected although we have not been able to find any short proof of this. The simplest one we have gotten is given in Appendix D. It is complicated and long, and for that reason has been placed in an Appendix. The reader is advised to skip the proof as well as the proofs of later theorems in a first reading.

Let us now turn to another similarity group, $S = GL(2)$. This case is easier, and we first make some preliminary remarks.

Definition. For a simple closed m-polygon with vertices $\{v_0, v_1, \ldots, v_{m-1}\}$, let e_j be the directed line segment $\overline{v_{j-1}, v_j}$. The set $\{e_1, e_2, \ldots, e_m\}$ will be referred to as the *edges*. The term diagonal will denote directed line segments between vertices which are not edges. Without loss of generality the simple closed polygons are from now on assumed to have a positive (counterclockwise) orientation. The interior of the polygon is the simply connected region with compact closure to the left of the polygon.

Lemma 2.1. *For every simple, closed m-polygons, $m \leq 4$, the closure of its interior can be decomposed into triangles by means of diagonals lying* *in the interior of the polygon.*

Proof: See Lemma 2, p. 17, K. Knopp (1945) or N.J. Lennes (1911).

Lemma 2.2. *For the triangularization created in Lemma 2.1 via internal diagonals, there exists at least one triangle with two sides being edges.*

Remark. It should be noted that this depends on the simply connected nature of the interior of a simple closed curve. For example, the annular region in Figure 2 is triangulated with all triangles having one edge.

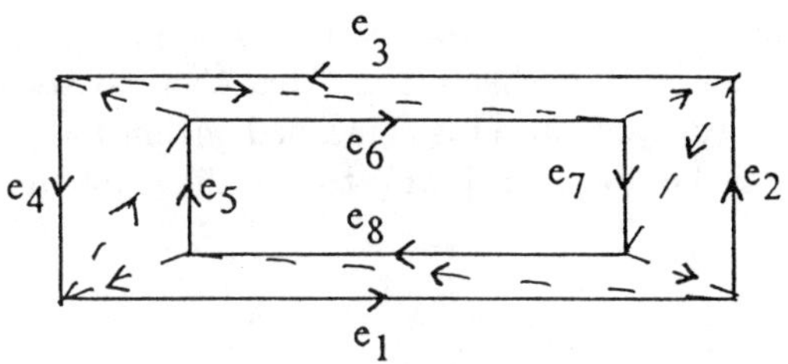

FIGURE 2

Proof (of Lemma 2.2):

Suppose that there exists a vertex v_j, with no diagonal involving v_j. Both e_j and e_{j+1} must be sides of triangles of the triangularization. The other two sides of the triangle involving e_j must join e_j at its endpoints, v_{j-1} and v_j. Consequently e_{j+1} must be the side joining e_j at v_j and the proof is done. Therefore we can assume that there exists at least one diagonal joined to each vertex. The remainder of the proof is by contradiction. Assume the conclusion is not true and using finite induction we will show that for any vertex v_k, v_k cannot be connected to either v_{k-s} or v_{k+s} by a diagonal, $1 \leq s \leq [\frac{m}{2}], 0 \leq k \leq m-1$, mod m addition. For s equal to one, each v_k is already connected by edges to v_{k-1} and v_{k+1} and consequently any diagonal would not be internal. Assume that we have shown it to be true for s. If for some k, v_k were connected to v_{k+s+1} (same argument for v_{k-s-1}) then we have the interior of the simple closed subpolygon $\{e_{k+1}, \dots e_{k+s+1},$ diagonal$\}$ contained within the interior of the original polygon. Each of the diagonals from $v_{k+1}, v_{k+2}, \dots, v_{k+s}$ must lie within this subpolygon and hence must connect to vertices from $\{v_k, v_{k+1}, \dots, v_{k+s}, v_{k+s+1}\}$. However, the maximum difference between indices from the two sets is $s, s \geq 1$, and thus by assumption there are no diagonals within the subpolygon; this contradicts v_{k+1} having a diagonal. Hence the statement is true for $s+1$. Therefore we have shown that no vertices are connected by diagonals. However, by Lemma 2.1 we know that for any simple closed polygon, $m \geq 4$, there exists an internal diagonal.

Q.E.D.

Theorem 2.1. *Any two given simple, closed m-polygons can be deformed into one another via a finite sequence of simple closed m-polygons. That is, $\mathcal{C}(\mathcal{R})(\subset \boldsymbol{R}^{2m})$ is polygonally-connected.*

Proof: First consider the convex case. The proof is by induction. For m equal to 3 it is easily seen that the result is true. Consequently we now assume it true for p and need to show it true for $p+1$. By convexity we can "push down" a given vertex, v_j, in a finite number of steps onto the line segment $\overline{v_{j-1}, v_{j+1}}$, replacing the edges e_j and e_{j+1} with this line segment, and thus reduce to the case of convex p-polygons for which the result is assumed true. Consequently the result is true for convex polygons. The general case is by an analogous induction argument since for a given $(p+1)$-polygon we know by Lemma 2.2 that there exists a vertex v_j such that the triangle (v_{j-1}, v_j, v_{j+1}) is contained within the polygon and thus the vertex v_j can be "pushed down" to the diagonal thus reducing to a p-polygon.

Q.E.D.

We shall use still another similarity group in the computer experiment, namely $S = US(2) \times 0(2)$. But we can achieve the same continuous deformations of polygons with this S as with $GL(2)$. Therefore we can state that connectivity holds also for this choice of S.

Once we know that connectivity holds, so that for any $c', c'' \in \mathcal{C}(\mathcal{R})$ there is a continuous path Γ in the interior of $\mathcal{C}(\mathcal{R})$ connecting c' with c'', we can find a path Γ_{spiral} that represents updatings using prescribed sweep areas. Indeed, we can approximate Γ to any accuracy we want by a "spiral" Γ_{spiral} made up of arcs a_ν. Here a_ν has all coordinates constant except those belonging to one of the sweep areas. This means that we can get from c' to c'' by a sequence of updatings all of which take place in $\mathcal{C}(\mathcal{R})$.

To get a better feeling for what is the effect of a local update let us consider the situation when the sweep area is of size one – it consists of single site i, and when $G = \boldsymbol{R}$. Given the values $u = g_{i-1}, w = g_{i+1}$ of the neighboring generators the conditional density for $v = g_i$ is

$$p(v) \propto A(u, v)Q(v)A(v, w). \tag{2.2}$$

To understand the influence of the acceptor function on an individual updating let us consider the conditional density for $x = g$

$$p(x, u, v) \propto A(x-u)A(v-x), \ \textit{say}\ u < v, \tag{2.3}$$

where for simplicity we have made the weight function uniform, $Q \equiv 1$. In (2.3) u, x, and v are the respective values at sites $i-1, i, i+1$.

Writing $A(t) = exp\ a(t)$, let us assume, as will usually be the case, that a(t) is an even function, decreasing for $t > 0$. Let us separate three cases

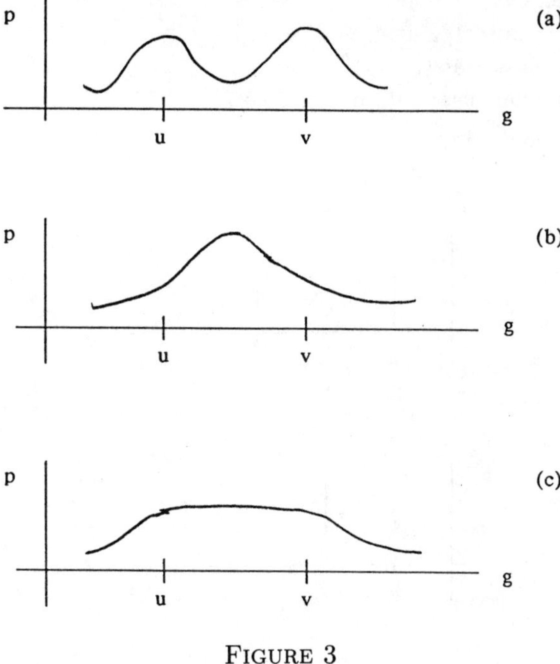

FIGURE 3

to be illustrated in Figure 3. In all three it is clear that p decreases toward zero as we move away from the interval (u, v).

If $a(t)$ is convex for $t > 0$ it follows that $a(x-u)+a(v-x) = a(x-u)+a(x-v)$ is convex in (u, v) and p looks like (a) in the figure, it is bimodal.

On the other hand if $a(t)$ is concave for $t > 0$ we get the unimodal behavior shown in (b).

An intermediate case is when $a(t) = c(t)$ which results in the flat density exhibited in (c).

The three cases are qualitatively very different, and to separate them we shall denote them as the

$$\left.\begin{array}{l}\text{alternative}\\ \text{compromising}\\ \text{neutral}\end{array}\right\}\text{behavior}$$

These terms will be obvious when we consider the special case $a(t) = -|t/\epsilon|^p$ where we have inserted a coupling parameter ϵ. It is logarithmically concave for $p > 1$. For $p = 1$ we get the intermediate case. Therefore, the stochastic updating of x, given the environment u and v, will tend to choose values close to alternatively u or v for $p < 1$, but compromise and choose a value close to the average of u and v if $p > 1$. For $p = 1$ it is neutral. This behavior becomes pronounced as $\epsilon \downarrow 0$. The limiting updating distributions are then given by

(a) point masses 1/2 at each of u and v
(b) point mass 1 at $(u+v)/2$
(c) rectangular distribution over (u,v)

as shown in Figure 4

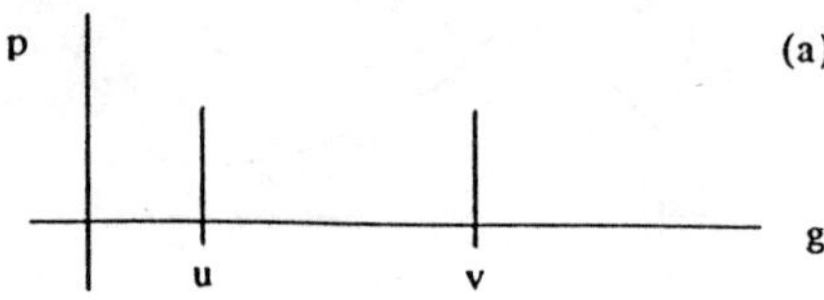

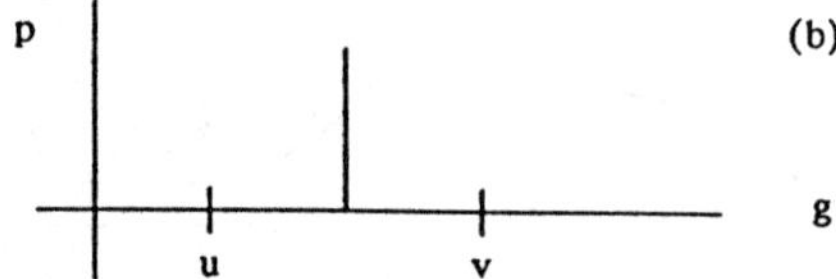

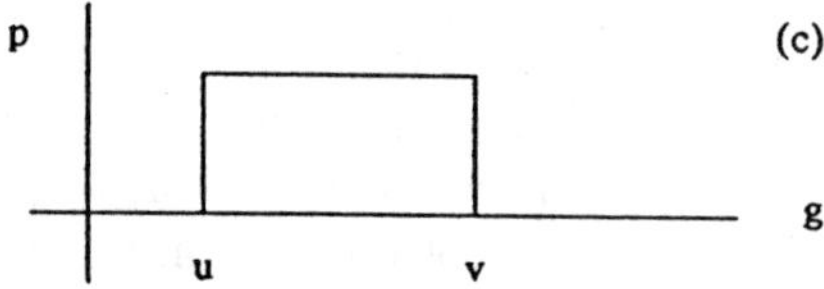

FIGURE 4

The following argument and lemma make precise the above for the case of $A(\cdot)$ being log concave.

Consider the following density of a Gibbs measure on $\boldsymbol{R}_+^n = \{(x_0, x_1, \ldots, x_{n-1}) | x_i^{'s} > 0\}$:

$$p(\ell_0, \ell_1, \ldots \ell_{n-1}) = \frac{1}{z} \prod_{j=0}^{n-1} A(\ell_j, \ell_{j+1}) \prod_{j=0}^{n-1} Q(\ell_j) (\text{mod n addition}) \tag{2.4}$$

and

$$\begin{aligned} p(\ell_j | \ell_k^{'s}, k \neq j) &= p(\ell_j | \ell_{j-1}, \ell_{j+1}) \\ &= \frac{1}{z} A(\ell_j, \ell_{j+1}) A(\ell_{j-1}, \ell_j) Q(\ell_j) \end{aligned} \tag{2.5}$$

where z and $\overline{z}$ are normalizing constants. Later in this section we will restrict the measures to the set $\mathcal{P}_n(\theta_0, \ldots, \theta_{n-1})$ for a fixed $(\theta_0, \ldots, \theta_{n-1})$. Assume that $A(\ell_{j-1}, \ell_j) = e^{a(\ell_{j-1}-\ell_j)}$ and $Q(\ell_j) = e^{q(\ell_j)}$ where a and q are $\mathcal{C}^1$ concave functions, a being even and q having at maximum at M. Let $\{\underline{X}_t\}$ be the temporally homogeneous Markov chain in discrete time with state space $\boldsymbol{R}^n_+$; $\underline{X}_0$ is a fixed point in $\boldsymbol{R}^n_+$. For simplicity of notation let ℓ_j be denoted by y with $x_1 = Min\{\ell_{j-1}, \ell_{j+1}\}$ and $x_2 = Max\{\ell_{j-1}, \ell_{j+1}\}$ and $\overline{x} = (\ell_{j-1} + \ell_{j+1})/2$. Consequently,

$$\frac{d\ log\ p(y)}{dy} = \frac{p'(y)}{p(y)} = a'(y - x_1) + a'(y - x_2) + q'(y) \tag{2.6}$$

is a concave function on $\boldsymbol{R}$ with

$$\frac{p'(M)}{p(M)} = a'(M - x_1) + a'(M - x_2) > 0 \tag{2.7}$$

$$\frac{p'(x_2)}{p(x_2)} = a'(x_2 - x_1) + q'(x_2) < 0. \tag{2.8}$$

By concavity we have that

$$\frac{p'(y)}{p(y)} < \frac{p'(\overline{x})}{p(\overline{x})} < 0 \qquad\qquad for\ y > \overline{x} \tag{2.9}$$

$$\frac{p'(y)}{p(y)} > \frac{p'(M)}{p(M)} > 0 \qquad\qquad for\ y < M \tag{2.10}$$

so that upon integrating $\frac{d \log\ p}{dy}$ we obtain for $u > 0$:

$$p(u + \overline{x}) \leq p(\overline{x})e^{(q'(\overline{x}))u}$$
$$p(M - u) \leq p(M)e^{-(a'(M-x_1)+a'(M-x_2))u}$$

Using these bounds we have

$$\int_{\overline{x}}^{\infty} p(y)dy \leq \int_0^{\infty} \leq \frac{p(\overline{x})}{-q'(\overline{x})} = o(\overline{x})$$

and

$$\int_0^M p(y)dy = \int_0^{\infty} p(M - u)du \leq \frac{p(M)}{(a'(M - x_1) + a'(M - x_2)} = o(\overline{x}).$$

Similarly one obtains

$$\int_{\overline{x}}^{\infty} yp(y)dy = o(\overline{x}) \text{ and } \int_{0}^{M} yp(y)dy = o(\overline{x})$$

For example, the second integral is:

$$\begin{aligned}
|\int_0^M yp(y)dy| &\leq |M\int_0^\infty p(M-u)du - \int_0^\infty up(M-u)du| \\
&\leq \frac{Mp(M)}{[a'(M-x_1)+a'(M-x_2)]} + p(M)\int_0^\infty ue^{-(a'(M-x_1)+a'(M-x_2))u}du \\
&\leq \frac{p(M)}{[a'(M-x_1)+a'(M-x_2)]}\left(M + \frac{1}{a'(M-x_1)+a'(M-x_2))}\right) \\
&= o(\overline{x})
\end{aligned}$$

Consequently, to evaluate

$$\begin{aligned}
E[\ell_j^{t+1}|\ell_{j-1}^t, \ell_{j+1}^t] &= \int_{-infty}^{\infty} yp(y)dy/\int_{-\infty}^{\infty} p(y)dy \\
&= \left[\int_M^{\overline{x}} yp(y)dy/\int_M^{\overline{x}} p(y)dy\right] + o(\overline{x})
\end{aligned} \tag{2.11}$$

we can restrict ourselves to the interval $[M, \overline{x}]$ where x is large enough.

In Section 2.1 we showed that $\{X_t\}$ is ergodic. The following lemma reveals the behavior of the local updating; intuitively, for ergodicity to occur large values must tend to be pulled back. Tweedie (1975) contains a result (Theorem 6.1) similar in spirit to this lemma but whose conditions would be difficult to verify in the present situation; Tweedie (1975) also gives an example to show that the ϵ (in the lemma below) must be uniform.

Lemma 2.3. *Assume that a and q are $\mathcal{C}^1$ concave functions with q having a maximum at M, a being an even function about zero, $a'(x) \to \infty$ as $x \to \infty, a''(x) = 0(1)$. Under these conditions, there exists a $B > 0, k > 0$ and $\epsilon > 0$ such that*

$$E\left[\ell_j^{t+1}|\ell_{j-1}^{(t)}, \ell_{j+1}^{(t)}\right]_{\leq\ B\ if\ |\overline{x}|\leq k}^{\leq\ \overline{x}-\epsilon\ if\ |\overline{x}|>k}$$

Proof: Let $m(\overline{x})$ be the mode of $p(\cdot)\chi_{[M,\overline{x}]}/\int_M^{\overline{x}} p(y)dy$. Then since $p(\cdot)$ is $\mathcal{C}^1$,the function $m(\cdot)$ satisfies the functional equation

$$a'(m(\overline{x})-x_1) + a'(m(\overline{x})-x_2) + q'(m(\overline{x})) = 0$$

or equivalently, using the evenness of a,

$$\begin{aligned}
a'\left(\left(\frac{x_2-x_1}{2}\right) + (m(\overline{x})-\overline{x})\right) - a'\left(\left(\frac{x_2-x_1}{2}\right) - (m(\overline{x})-\overline{x})\right) \\
+ q'(m(\overline{x})) = 0.
\end{aligned} \tag{2.12}$$

We wish to show that $\exists \epsilon > 0 \ni$

$$\sup_{\overline{x}>B}(m(\overline{x}) - \overline{x}) < -\epsilon \tag{2.13}$$

where B is sufficiently large. Suppose this were not the case, that is that $m(\overline{x}) - \overline{x} \to 0$ as $x \to \infty$. If this occurs, then rewriting expression (2.12) as:

(2.14)

$$\left[\frac{a'\left(\left(\frac{x_2-x_1}{2}\right) + (m(\overline{x}) - \overline{x})\right) - a'\left(\left(\frac{x_2-x_1}{2}\right) - (m(\overline{x}) - \overline{x})\right)}{(m(\overline{x}) - \overline{x})}\right] \times (m(\overline{x}) - \overline{x})$$
$$+ q'(m(\overline{x})) = 0$$

and using that $q'(m(\overline{x}))$ is negative for large $\overline{x}$, the first term must be bounded away from zero from above for large $\overline{x}$. However, if $x_2 - x_1$ is bounded (as $\overline{x} \to \infty$), then the first two terms in expression (2.12) go to zero. If $x_2 - x_1$ is unbounded then since $a''(z) = 0(1)$, the first term in expression (2.14) goes to zero. Therefore expression (2.13) is satisfied for some $\epsilon > 0$ and $k > 0$. By expressions (2.9) and (2.10) it follows that p is nondecreasing below the mode and nonincreasing above. Let $\lambda(\overline{x})$ be defined as:

$$\lambda(\overline{x}) = \int_M^{\overline{x}-\epsilon} p(y)dy \Big/ \int_n^{\overline{x}} p(y)dy$$

It follows that

$$E[\ell_j^{t+1}|\ell_{j+1}^t] = \left[\int_M^{\overline{x}} yp(y)dy / \int_M^{\overline{x}} p(y)dy\right] + o(x)$$
$$\leq \left[\left(\int_M^{\overline{x}-\epsilon} yp(y)dy + \int_{\overline{x}-\epsilon}^{\overline{x}} yp(y)dy\right) \Big/ \int_M^{\overline{x}} p(y)dy\right] + o(\overline{x})$$
$$\leq (\overline{x} - \epsilon)\lambda(\overline{x}) + (1 - \lambda/((\overline{x}))\left[\int_{\overline{x}-\epsilon}^{\overline{x}} yp(y)dy \Big/ \int_{\overline{x}-\epsilon}^{\overline{x}} p(y)dy\right] + o(\overline{x})$$

and since, $0 < \lambda(\overline{x}) < 1$, and the mode, $m(\overline{x})$, satisfies $m(\overline{x}) < \overline{x} - \epsilon$, for $\overline{x} > k$, we have that $p(\cdot)$ is nonincreasing on $[\overline{x} - \epsilon, \overline{x}]$ so that the mean of $p(\cdot)$ restricted to this interval and normalized must lie below the midpoint $\overline{x} - \frac{1}{2}\epsilon$. Therefore, we have

$$E[\ell_j^{t+1}|\ell_{j-1}^t, \ell_{j+1}^t] \leq \lambda(\overline{x})(\overline{x} - \epsilon) + (1 - \lambda(\overline{x}))(\overline{x} - \frac{1}{2}\epsilon) \leq \overline{x} - \frac{1}{2}\ \epsilon.$$

Q.E.D.

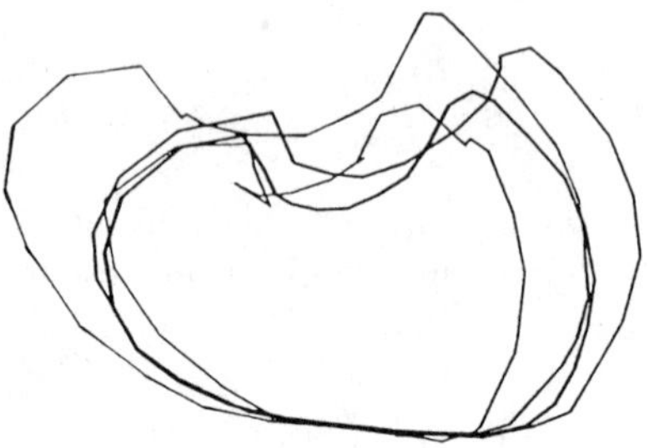

FIGURE 5

We illustrate the application of stochastic relaxation to image synthesis by the picture ensemble shown in Figure 5. Here the template is a 2-dimension projection of a human stomach taken from Anson (1963), Q is an exponential function, and we made the rather extreme choice $A \equiv 1$.

From this and other similar preliminary experiments we learned the following.

a) Having $A \equiv 1$ does not lead to pictures of the sort we want. It implies that the s-values are independent except for the (weak) dependence caused by the global constraint. It makes the *boundaries too irregular.*

b) Another effect is to generate *self intersection of a local nature* too often: the curve intersects itself at close points.

c) Increasing the couplings by a better choice of acceptor function we get better looking picture ensembles (many examples will appear in section 4), and self intersection becomes very rare.

§2.3. Stochastic relaxation can, in principle, be applied to any pattern theoretic model for synthesis. This *universality* is counteracted, however, by the fact that it sometimes *requires massive computing power*, especially when n, the size of the configuration, is large or when couplings are strong.

Choosing the sweep strategy appropriately we can implement stochastic relaxation in parallel. When parallel computer architecture becomes generally available this will expand the applicability of the method.

There is, however, *another way to handle the computational problem, namely by analytical means.* A good deal of attention has been given to establish *limit theorems in metric pattern theory*; see Grenander (1981), Chapter 5 for a general discussion of this topic, and Chow-Grenander (1985) for details. The following modification of a recent result from Chow (1987) has important consequences for the current study.

We start by introducing some notation. Let $(U_0, U_1, \ldots, U_{n-1})$ be a random sample from a distribution with density

$$p(u_0, u_1, \ldots, u_{n-1}) = \left[\prod_{j=0}^{n-1} A\left(\frac{u_{i+1} - u_i}{\epsilon_n}\right) \cdot Q(u_j)\right] \bigg/ Z_n, \ (u_n = u_0)$$

where A is the acceptance function and Z_n a normalizing constant. Rescale U by introducing $(X_0, X_1, \ldots, X_{n-1}) = \sqrt{\epsilon_n}(U_0, U_1, \ldots, U_{n-1})$. To study the asymptotic behavior of $(X_0, X_1, \ldots, X_{n-1})$ as a whole, we define

$$X_n(t) = X_{[nt]} + n(t - [nt]/n)(X_{[nt]} - X_{[nt]}), \ 0 \le t \le 1.$$

Here we require $x_i = x_j$ for $i \equiv j$ (mod n). Hence $X_n(\cdot) \in C[0,1]$ and $X_n(0) = X_n(1)$.

Assume ϵ_n and the nonnegative functions A and Q satisfy
(A1) $\epsilon_n = c^2/n$, where $c > 0$ is a constant.
(A2) $\int Q(x)dx < \infty, \int A(x)dx = 1, \int xA(x)dx = 0$ and $\int x^2 A(x)dx = \sigma^2 < \infty$.
(A3) $A(\cdot)$ is bounded on $(-\infty, \infty)$.
(A4) For any $\delta > 0$, $\sup_{|x| \ge \delta} Q(x) < Q(0)$ and $Q''(0) < 0$.

For clarity of expression we shall always write $Q(x) = Q(0)exp(-q(x))$. It is easy to check that (A4) is equivalent to:
(A4) For any $\delta > 0, \inf_{|x| \ge \delta} q(x) > 0$ and $q(x) = 2^{-1}r^2x^2 + o(x^2)$ for x small, where $r = (-Q''(0)/Q(0))^{1/2} > 0$.

Then we have the following result for the cyclic connector graph.

Theorem 2.2. *$\{X_n(\cdot)\}$ converges weakly in $C[0,1]$ to a stationary, Gaussian Markov process $\{X(\cdot)\}$ with mean 0 and covariance*

$$R(s,t) = \frac{\sigma \cdot cosh[\frac{1}{2} - |s-t|)r\sigma c^2]}{2r \cdot sinh(\frac{1}{2}r\sigma c^2)}, \ 0 \le s,t \le 1, \tag{2.15}$$

where $r = (-Q''(0)/Q(0))^{1/2} > 0$.

The proof, which is technical and lengthy, follows a standard procedure and is divided into three steps. Note that $(X_0, X_1, \ldots, X_{n-1})$ has a density on $\boldsymbol{R}^n$ proportional to

$$\left[\prod_{j=0}^{n-1} A((x_{j+1} - x_j)/\sqrt{\epsilon_n})\right] \cdot exp\left(-\sum_{j=0}^{n-1} q(\sqrt{\epsilon_n}x_j)\right), \ x_n = x_0 \tag{2.16}$$

Step 1. Show that $\{X_n(\cdot)\}$ converges weakly in $C[0,1]$ and an invariance principle holds. That is, the limit distribution depends only on c, σ and r for all functions A, Q satisfying (A1)-(A4). This is the crucial step. Once it is done, the remaining two steps are straightforward and easy.

Step 2. Find the limit distribution by choosing $A(x) = \frac{1}{\sqrt{2\pi\sigma^2}} exp\left(-\frac{x^2}{2\sigma^2}\right)$ and $Q(x) = exp(-\frac{r^2x^2}{2})$. In this special case everything is Gaussian. Direct computation of the covariance matrix can be made.

Proof of Step 2. It is clear from (2.16) that $(X_n(n^{-1}), X_n(2n^{-1}), \ldots, X_n(1))$ has a joint Gaussian distribution with mean 0 and their covariance matrix $R_n = (EX_n(i/n)X_n(j/n))_{n\times n}$ satisfying

$$(R_n)^{-1} = n(\sigma c)^{-2} M_n(r^2\sigma^2c^4/n^2) \tag{2.17}$$

where in general $M_n(h), n \geq 1, h > 0$, is an $n \times n$ circulant matrix defined by

$$(M_n(h))_{ij} = \begin{cases} 2+h, & if\ i = j \\ -1, & if\ i-j = \pm 1 (mod\ n) \\ 0, & otherwise. \end{cases}$$

It is well-known [Davis (1979)] that the eigenvalues of $M_n(h)$ are $2+h-2cos(2\pi(k-1)/n), k = 1, 2, \ldots, n$, and

$$(M_n(h))^{-1} = \sum_{k=1}^{n} B_k(h)/[2+h-2cos(2\pi(k-1)/n)] \tag{2.18}$$

where $(B_k(h))_{uv} = n^{-1}(exp - i2\pi/n)^{(u-v)(k-1)}$.

For $0 \leq s \leq t \leq 1$, we have by (2.17) and (2.18),

$$E(X_n([sn]/n)X_n([tn]/n)) = \frac{\sigma^2c^2}{n^2} \sum_{k=1}^{n} \frac{exp\{i2\pi([nt]-[ns])(k-1)/n\}}{2(1-cos(2\pi(k-1)/n)) + r^2\sigma^2c^4/n^2}.$$

Using the fact that the left-hand side of the previous equation is real and then changing the summation from $-[n/2]$ to $[(n-1)/n]$

$$E(X_n([sn]/n)X_n([tn]/n)) = \frac{\sigma^2c^2}{n^2} \sum_{-[n/2]}^{[(n-1)/2]} \frac{cos(2\pi([nt]-[ns])(k-1)/n)}{2(1-cos(2\pi(k-1)/n)) + r^2\sigma^2c^4/n^2}.$$

Because $(x^2/2) - (x^4/4!) \leq 1 - cosx \leq x^2/2$ it is not hard to show by truncation and bounded convergence theorem that for $0 \leq s \leq t \leq 1$

$$\begin{aligned} \lim_n E(X_n([sn]/n)X_n([tn]/n)) &= \\ &= \sigma^2c^2 \sum_{-\infty}^{\infty} [cos(2\pi(t-s)k)](4\pi^2k^2 + r^2\sigma^2c^4)^{-1} \\ &= \sum_{-\infty}^{\infty} (-1)^k [cos(\pi k(1+2(s-t)))](4\pi^2k^2 + r^2\sigma^2c^4)^{-1} \\ &= \frac{\sigma cosh[(2^{-1}+(s-t))r\sigma c^2]}{2r\ sinh(2^{-1}r\sigma c^2)} \end{aligned}$$

where in the last equality we have used the following formula [Mitrinović and Keckić (1984]: for α, β and $t \in \boldsymbol{R}$ with $|\alpha| < |\beta|$

$$\sum_{-\infty}^{\infty} \frac{(-1)^k \beta t}{\pi^2 k^2 + \beta^2 t^2} cos\left(\frac{\pi \alpha k}{\beta}\right) = \frac{cosh(\alpha t)}{sinh(\beta t)}.$$

This proves (2.15).

It is clear from (2.15) that $\{X(\cdot)\}$ is stationary. Since (2.16) implies that $\{X_n(\cdot)\}$ is Markovian, it is reasonable to expect the limiting process $\{M(\cdot)\}$ to be Markovian. In fact this can be checked directly by using the stationary and Gaussian property of $\{X(\cdot)\}$ for $0 < u < v < 1$.

Step 1 will be achieved through several lemmas. Let P_n be the probability measure on $\boldsymbol{R} \times C[0,1]$ induced by the mapping which sends $(X_0, X_1, \ldots, X_{n-1})$ to $(X_0, X_n(\cdot) - X_0)$. Since $X_n(t) = X_0 + (X_n(t) - X_0)$ and the mapping $T : \boldsymbol{R} \times C[0,1] \to C[0,1]$ defined by

$$T(x_0, x(\cdot)) = x_0 + x(\cdot) \tag{2.19}$$

is continuous, the limiting behavior of the stochastic process $\{X_n(\cdot)\}$ can be determined [Billingsley (1968), p.30] from that of P_n.

A reference measure is needed for the study of $X_n(\cdot) - X_0$. Let $(V_1, V_2, \ldots, V_n)$ be a random sample from a distribution on R^n with density

$$(\sqrt{\epsilon_n})^{-n} \prod_{i=0}^{n-1} A((V_{i+1} - V_i)/\sqrt{\epsilon_n}), \ V_0 \equiv 0 \tag{2.20}$$

and let B_n be the probability measure on $C[0,1]$ induced by the mapping which maps $(V_1, V_2, \ldots, V_n)$ to

$$V_n(t) = V_{[nt]} + n(t - [nt]/n)(V[nt] + 1 - V_{[nt]}), \ 0 \le t \le 1$$

Similarly, let $(\tilde{V}_1, \tilde{V}_2, \ldots, \tilde{V}_{n-1})$ be a random sample from a distribution on $\boldsymbol{R}^{n-1}$ with its density proportional to

$$\prod_{i=0}^{n-1} A((V_{i+1} - V_i)/\sqrt{\epsilon_n}), \ V_0 = V_n = 0 \tag{2.21}$$

and let W_n be the measure on $C[0,1]$ induced by the mapping which maps $(\tilde{V}_1, \tilde{V}_2, \ldots, \tilde{V}_{n-1})$ to

$$\tilde{V}_n(t) = \tilde{V}_{[nt]} + n(t - [nt]/n)(\tilde{V}_{[nt]+1} - \tilde{V}_{[nt]}), \tilde{V}_0 = \tilde{V}_n = 0, 0 \le t \le 1.$$

Note that

$$W_n \stackrel{\mathcal{D}}{=} B_{n_{|x(1)=0}} \tag{2.22}$$

It is clear from (2.21) that $(V_{i+1}-V_i)/\sqrt{\epsilon_n}, 0 \le i \le n-1 (V_0 \equiv 0)$, are i.i.d. random variables with density A. Since

$$V_i = \sqrt{\epsilon_n}\left[\sum_{j=0}^{i-1}(V_{j+1}-V_j)/\sqrt{\epsilon_n}\right] = c\left[\sum_{j=0}^{i-1}(V_{j+1}-V_j)/\sqrt{\epsilon_n}\right]/\sqrt{n},$$

we have from Donsker's invariance principle [Billingsley (1968)], that $B_n \xrightarrow{n} c\sigma W$, where W is the standard Wiener measure on $C[0,1]$. Because $W_{|x(1)=0} \overset{\mathcal{D}}{=} W_0$, the Brownian bridge, it is natural to guess from (2.22) that $W_n \xrightarrow{n} c\sigma W_0$. In fact this is true and can be proved by appealing to Theorem 3.1 in Wichura (1968). The details can be found in Chow (1987) and the proof is omitted.

Lemma 2.4. *Under the assumptions (A1)-(A3), $W_n \xrightarrow{n} c\sigma W_0$, where W_0 is the Brownian bridge.*

Write $Y_i = X_i - X_0$ for $1 \le i \le n$. By (2.16), $(X_0, Y_1, \ldots, Y_{n-1})$ has a density on $\mathcal{R}^n$ which is proportional to

(2.23)

$$\prod_{i=0}^{n-1} A((y_{i+1}-y_i)/\sqrt{\epsilon_n}) \cdot (exp - [q(\sqrt{\epsilon_n}x_0) + \sum_{i=1}^{n-1} q(\sqrt{\epsilon_n}(y_i + x_0))]), y_0 \equiv 0.$$

Note that $X_n(0) - X_0 = 0$ and $X_n(\cdot) - X_0$ can be obtained by joining $Y_0 = 0, Y_1, \ldots, Y_n$ at $t = 0, n^{-1}, \ldots, 1$.

Let μ_n be the marginal distribution of X_0 and $P_n(\cdot|x_0)$ be the distribution of $X_n(\cdot) - X_0$ on $C[0,1]$ conditioned on $X_0 = x_0$. Then

$$P_n(B \times E) = \int_B P_n(E|x_0)d\mu_n(x_0) \tag{2.24}$$

holds for each $B \in B(\boldsymbol{R})$, the Borel σ-field on $\boldsymbol{R}$, and each $E \in B(C[0,1])$, the σ-field on $C[0,1]$ generated by the finite dimensional cylinder sets.

Our approach is to study the limiting behavior of $P_n(\cdot|x_0)$ and μ_n. Then that of P_n follows immediately. By using (2.21) and (2.23) it can be shown easily that

Lemma 2.5. *(i) For fixed $x_0 \in \boldsymbol{R}$, $P_n(\cdot|x_0)$ is absolutely continuous with respect to W_n and*

$$dP_n(x(\cdot)|x_0)/dW_n = [exp - \sum_{j=1}^{n-1} q(\sqrt{\epsilon_n}(x(i/n) + x_0))]/D_n(x_0) \tag{2.25}$$

where

$$D_n(x_0) = \int_{C[0,1]} [exp - \sum_{j=1}^{n-1} q(\sqrt{\epsilon_n}(x(i/n) + x_0))]dW_n.$$

(ii) μ_n has a density (with respect to Lebesgue measure on $\boldsymbol{R}$) g_n with

$$g_n(x_0) = D_n(x_0)[exp - q(\sqrt{\epsilon_n}x_0)]/G_n \tag{2.26}$$

where

$$G_n = \int D_n(x_0)[exp - q(\sqrt{\epsilon_n}x_0)]dx_0.$$

Because $W_n \to c\sigma W_0$, the behavior of $x(\cdot)$ under W_n is well understood. Under the assumptions (A1) and (A4)

$$\sum_{i=1}^{n-1} q(\sqrt{\epsilon_n}(x(i/n)+x_0)) \approx 2^{-1}r^2\epsilon_n(x(i/n)+x_0)^2$$
$$\to 2^{-1}r^2c^2\int_0^1 (x(t)+x_0)^2dt.$$

this makes the following lemma plausible.

Lemma 2.6. *(i) For fixed $x_0 \in \boldsymbol{R}$, $P_n(\cdot|x_0)$ converges weakly to a measure $P(\cdot|x_0)$ which is absolute continuous with respect to $c\sigma W_0$, and*

$$dP_n(x(\cdot|x_0)/d(c\sigma W_0) = [exp - 2^{-1}r^2c^2\int_0^1 (x(t)+x_0)^2dt]/D_n(x_0) \tag{2.27}$$

where, with $b^2 = [\sigma \cdot coth(r\sigma c^2/2)]/(2r)$,
(2.28)

$$D(x_0) = lim\ D_n(x_0) = EW_0\{exp - \left[2^{-1}r^2c^2\int_0^1 (c\sigma x(t)+x_0)^2dt\right]\}$$
$$= (r\sigma c^2/sinh(r\sigma c^2)).$$

(ii) $\lim_n g_n(x_0) = g(x_0) \equiv (2\pi b^2)^{-1/2}exp[-(2b^2)^{-1}x_0^2]$.

Assume tentatively that Lemma 2.6 holds. Since all g_n and g are density functions on $\boldsymbol{R}$, Scheffe's theorem [Billingsley (1968)] implies

$$lim\int |g_n(x_0)-g(x_0)|dx_0 = 0. \tag{2.29}$$

Because $0 \leq P(E|x_0) \leq 1$, it follows from (2.25),(2.29) and Lemma 2.6 that

$$\lim_n P_n(B\times E) = \int_B P(E|x_0)g(x_0)dx_0 \tag{2.30}$$

holds for each $B \in \mathcal{B}(\boldsymbol{R})$ and each $E \in \mathcal{B}(C[0,1])$ with $(c\sigma W_0)(\partial E) = 0$.

Since the right-hand side of (2.30) defines a measure on $\boldsymbol{R}\times C[0,1]$, we have shown [Billingsley (1968); Theorem 3.1] that P_n converges weakly.

Then by the continuity of the mapping T defined in (2.19), and in view of Lemma 2.6, $\{X_n(\cdot)\}$ has a weak limit in $C[0,1]$ which depends only on c, σ and r. Step 1, and thus Theorem 2.5, are proved, except that we have to verify Lemma 2.6.

Proof of Lemma 2.6. (i) For simplicity of notations, let $F_n(x;x_0) = \sum_{i=1}^{n-1} q(\sqrt{\epsilon_n}(x(i/n)+x_0))$ and $F_n(x;x_0) = 2^{-1}r^2c^2\int_0^1 (x(t)+x_0)^2dt$.

Suppose that we know already that for any $\epsilon > 0$

$$\lim_n W_n\{x \in C[0,1]:\ |F_n(x;x_0) - F(x;x_0)| > \epsilon\} = 0. \tag{2.31}$$

Observing that exponential function $exp(-s)$ is uniformly continuous and bounded on $[0,\infty)$,

$$\lim_n \int_{C[0,1]} |[exp - F_n(x;x_0)] - [exp - F(x;x_0)]| dW_n = 0. \tag{2.32}$$

Because $F_n(x;x_0)$ is a continuous function on $C[0,1]$ and $W_n \to c\sigma W_0$,

$$\begin{aligned} \lim_n \int_{C[0,1]} [exp - F(x;x_0)]dW_n &= \int_{C[0,1]} [exp - F(x;x_0)]d(c\sigma W_0) \\ &= \int_{C[0,1]} [exp - F(x;x_0)]dW_0 \end{aligned} \tag{2.33}$$

where the last equation is obtained by making a scale change.

By (2.30), (2.31) and the triangular inequality,

$$\lim_n \int_{C[0,1]} [exp - F_n(x;x_0)]dW_n = \int_{C[0,1]} [exp - F(c\sigma x;x_0)dW_0. \tag{2.34}$$

This proves (2.28) if we note that for $\xi, \eta \in \boldsymbol{R}$

(2.35)

$$E_{W_0}[exp - [2^{-1}\xi^2 \int_0^1 (x(t)+n\eta)^2dt]] = \sqrt{\frac{\xi}{sinh\ \xi}}exp\left[-\frac{\eta^2}{2}(2\xi\ tanh\frac{\xi}{2})\right],$$

which is an immediate consequence by letting $y(t) \equiv n$ in the following formula [Chiang, Chow and Lee (1986)]: for $\xi \in \boldsymbol{R}$ and $y(\cdot) \in L^2[0,1]$,

$$\begin{aligned} &E_{W_0}[exp - [2^{-1}\xi^2 \int_0^1 (x(t)+y(t))^2dt]] \\ &= \sqrt{\frac{xi}{sinh\ \xi}}exp\left[-\frac{\xi^2}{2}\left(\int_0^1\int_0^1 k(s,t;\xi)y(s)y(t)dsdt + \int_0^1 y^2(t)dt\right)\right] \end{aligned}$$

where $k(s,t;\xi) = \xi[cosh(\xi(1-|t+s|)) - cosh(\xi(1-|t-s|))]/(2\ sinh\ \xi)$.

Since (2.30) and (2.31) still hold when the integral is taken over a subset E of $C[0,1]$ with $(c\sigma W_0)(\partial E) = 0$, we have, similar to (2.34),

$$\lim_n \int_E [exp - F_n(x; x_0)] dW_n = \int_E [exp - F(x; x_0)] d(c\sigma W_0). \tag{2.36}$$

Now combine together (2.25), (2.34), (2.36) and (2.27),

$$\lim_n P_n(E|x_0) = P(E|x_0)$$

and part (i) is finished [Billingsley (1968); Theorem 2.11], except that it

$$\begin{aligned} &|n^{-1}\sum_{i=1}^{n}(x(i/n)+x_0)^2 - \int_0^1 (x(t)+x_0)^2 dt| \\ &= |\sum_{i=1}^{n}\int_{(i-1)/n}^{i/n} [(x(i/n)+x_0)^2 - (x(t)+x_0)^2] dt| \\ &\leq 2(d + |x_0|) w_x(n^{-1}). \end{aligned} \tag{2.37}$$

Hence, for $x \in H$,

$$\begin{aligned} |F_n(x; x_0) - F(x; x_0)| &= o(1) + 0(w_x(n^{-1})) \\ &\leq o(1) + 0(w_x(\delta)) \\ &\leq o(1) + 0(\epsilon_1) \leq \epsilon \end{aligned} \tag{2.38}$$

if ϵ_1 is properly chosen and n is large enough.

Because η in (2.37) can be arbitrarily small, (2.31) follows from (2.37) and (2.38).

(ii) It is clear from (2.26) and (2.28) that we need only to show

(2.39)

$$\begin{aligned} &lim\ G_n \\ &= (cr)^{-1}\sqrt{2\pi} E_{W_0}\left\{exp - \left[2^{-1}r^2\sigma^2c^4\left(\int_0^1 x^2(t)dt - \left(\int_0^1 x(t)dt\right)^2\right)\right]\right\} \\ &= (cr)^{-1}\sqrt{2\pi} r\sigma c^2 (2\ sinh\ (r\sigma c^2/2))^{-1}. \end{aligned}$$

Note that the last equality follows by integrating both sides of (2.35) over $\eta \in (-\infty, \infty)$.

Let $z = \sqrt{\epsilon_n} x_0 = cn^{-1/2} x_0$ to make a change of variable,

$$\begin{aligned} G_n &= \int D_n(x_0)[exp - q(\sqrt{\epsilon_n} x_0)] dx_0 \\ &= \int [exp - q(\sqrt{\epsilon_n} x_0)] dx_0 \int [exp - \sum_{i=1}^{n-1} q(\sqrt{\epsilon_n}(x(i/n) + x_0))] dW_n \\ &= c^{-1}\sqrt{n}\int [exp - q(z)] dz \int [exp - \sum_{i=1}^{n-1} q(z + dx(i/n)/\sqrt{n})] dW_n. \end{aligned}$$

We claim that for any $\epsilon > 0$

$$(2.40)\quad \sqrt{n}\int_{|z|\geq 2\epsilon}[exp - q(z)]dz\int[exp - \sum_{i=1}^{n-1} q(z+cx(i/n)/\sqrt{n})]dW_n \xrightarrow{n} 0$$

Denote it by I_n. Write $I_n = I_{1,n} + I_{2,n}$, where

$$(2.41)\quad \begin{aligned} I_{1,n} =&\sqrt{n}\int_{|z|\geq 2\epsilon}[exp - q(z)]dz\int_{\{x:\|x\|_\infty\leq\epsilon\sqrt{n}/c\}} \\ &[exp\sum_{i=1}^{n-1} q(z+cx(i/n)/\sqrt{n})]dW_n \\ I_{2,n} =&\sqrt{n}\int_{|z|\geq 2\epsilon}[exp - q(z)]dz\int_{\{x:\|x\|_\infty\leq\epsilon\sqrt{n}/c\}} \\ &[exp\sum_{i=1}^{n-1} q(z+cx(i/n)/\sqrt{n})]dW_n \end{aligned}$$

Observe that $|z+cx(i/n)/\sqrt{n}| \geq |z| - c\|x\|_\infty/\sqrt{n} \geq \epsilon$ in $I_{1,n}$. By (A4) $n \equiv \inf_{|u|\geq\epsilon} a(u) > 0$. Hence

$$(2.42)\quad \begin{aligned} I_{1,n} &\leq \sqrt{n}\int_{|z|\geq 2\epsilon}[exp - q(z)]dz\int_{\{x:\|x\|_\infty\leq\epsilon\sqrt{n}/c\}}[exp - n\eta]dW_n \\ &\leq \sqrt{n}[exp - n\eta]\int[exp - q(z)]dz \xrightarrow{n} 0. \end{aligned}$$

Suppose we know already that for any $\epsilon > 0$

$$(2.43)\quad W_n\{x\in C[0,1] : \|x\|_\infty > \epsilon\sqrt{n}/c\} = 0(n^{-1})$$

Because $0 \leq exp - q(\cdot) \leq 1$,

$$(2.44)\quad I_{2,n} \leq \sqrt{n}\{\int[exp - q(z)]dz\}\cdot W_n\{x : \|x\|_\infty > \epsilon\sqrt{n}/c\} = 0(1/\sqrt{n})$$

This, together with (2.42) implies (2.40). In fact we have shown in the proof of (2.42) and (2.44) that for any $\epsilon > 0$,

$$(2.45)\quad \begin{aligned} G_n =& \\ & c^{-1}\sqrt{n}\int_{|z|\geq 2\epsilon}[exp - q(z)]dz\int_{\{x:\|x\|_\infty\leq\epsilon\sqrt{n}/c\}} \\ &[exp - \sum_{i=1}^{n-1} q(z+cx(i/n)/\sqrt{n})]dW_n + o(1). \end{aligned}$$

When $|z| \leq 2\epsilon$ and $\|x\|_\infty \leq \sqrt{n}/c, |z + cx(i/n)/\sqrt{n}| \leq 3\epsilon$. If ϵ is chosen small enough, then by (A4) $[exp - q(z)] = 1 + o(1)$ and

$$\sum_{i=1}^{n-1} q(z + cx(i/n)/\sqrt{n}) = [2^{-1}r^2 + o(1)] \sum_{i=1}^{n-1} q(z + cx(i/n)/\sqrt{n})^2$$
$$[2^{-1}r^2 + o(1)][n(z + k_n(x))^2 + f_n(x)]$$

where $k_n(x) = cn^{-3/2} \sum_{i=1}^{n-1} x(i/n)$ and $f_n(x) = c^2 n^{-1} x^2(i/n) - (cn^{-1} - (cn^{-1} \sum_{i=1}^{n-1} x(i/n))^2$.

Changing the order of integration in (2.45)

$$G_n = \int_{\{x:\|x\|_\infty \leq \epsilon\sqrt{n}/c\}} [exp - (2^{-1}r^2 + o(1))f_n(x)]dW_n$$
$$\cdot c^{-1}\sqrt{n}(1 + o(1)) \int_{|z| \leq 2\epsilon} [exp - (2^{-1}r^2 + o(1))n(z + k_n(x))^2]dz + o(1).$$

Since $|k_n(x)| \leq \epsilon$ in the domain of integration, the integral with respect to z is bounded above that

$$\int_{|u| \leq 3\epsilon} [exp - (2^{-1}r^2 + o(1))nu^2]du = \sqrt{2\pi}/(\sqrt{n}r) + o(1).$$

and bounded below by

$$\int_{|u| \leq \epsilon} [exp - (2^{-1}r^2 + o(1))nu^2]du = \sqrt{2\pi}/(\sqrt{n}r) + o(1).$$

It follows that

$$G_n = (cr)^{-1}\sqrt{2\pi} \int_{\{x:\|x\|_\infty \leq \sqrt{n}/c\}} [exp - (2^{-1}r^2 + o(1))f_n(x)]dW_n + o(1).$$

Because $W_n \to c\sigma W_0$ and $lim\ f_n(x) = c^2 \int_0^1 x^2(t)dt - c^2 \left(\int_0^1 x(t)dt\right)^2$, (2.39) can be obtained easily. The details are omitted. Because the proof is similar to that of (2.45), except that we need to replace the domain of integration $\{x : \|x\|_\infty \leq \epsilon\sqrt{n}/c\}$ by $C[0,1]$ and $\{x : \|x\|_\infty \leq m\}$ to get upper and lower bounds and then let m tend to ∞.

It remains to check (2.43). let $\xi_1, \xi_2, \ldots$ is a sequence of i.i.d. random variables with density A. And let $S_j = \sum_{i=1}^{j} \xi_i$ and $T_j = \sum_{i=j+1}^{n} \xi_i$. (So $S_0 = T_n = 0$.) By using a lemma in [Freedman (1983); p.12]

$$\left[\min_{0 \leq j \leq n} P\{T_j \geq 0\}\right] P\left\{\max_{0 \leq j \leq n} S_j \geq C_n\right\} \leq P\{S_n \geq C_n\} \tag{2.46}$$

It is clear that $\min_{0 \le j \le n} P\{T_j \ge 0\} = \min_{0 \le j \le n} P\{S_j \ge 0\}$. Since $E\xi_i = \int xA(x)dx = 0$ and $E(\xi_i^2) = \int x^2 A(x)dx = \sigma^2, P\{S_j \ge 0\} > 0$ holds for each j and by Central Limit Theorem,

$$P\{S_j \ge 0\} = P\{S_j/(j\sigma^2)^{1/2} \ge 0\} \xrightarrow{j} 2^{-1}.$$

Hence, $\min_{0 \le j \le n} P\{T_j \ge 0\} \ge \min_{1 \le j} P\{S_j \ge 0\} \ge \delta > 0$ holds for some constant δ. Now applying Chebychev's inequality to the right-hand side of (2.46)

$$P\{\min_{0 \le j \le n} S_j \ge C_n\} = 0(1/n).$$

A similar result holds for $P\{\min_{0 \le j \le n} S_j \le -C_n\}$. Hence we have

$$P\{\max_{0 \le j \le n} |S_j| \ge C_n\} = 0(1/n). \tag{2.47}$$

By using (2.22) and the construction of B_n,

$$\begin{aligned} W_n\{x : \|x\|_\infty > \epsilon\sqrt{n}/c\} & \\ &= W_n(\{x : \|x\|_\infty > \epsilon\sqrt{n}/c\}|x(1) = 0) \\ &= P(\max_{1 \le j \le n} |S_j| \ge \epsilon n/c^2 | S_n = 0 \\ &\le 2P(\max_{1 \le j \le [(n+1)/2]} |S_j| \ge \epsilon n/c^2 | S_n = 0) \end{aligned} \tag{2.48}$$

because $\int uA(u)du = 0$ implies that S_j and S_{n-j} have the same distribution when conditioned on $S_n = 0$.

Define $\tau = inf\{j \le n : |S_j| \ge \epsilon n/c^2\}$. Then

$$\begin{aligned} P(&\max_{1 \le j \le [(n+1)/2]} |S_j| \ge \epsilon n/c^2 | S_n = 0) \\ &= \lim_{\delta \downarrow 0} P(\max_{1 \le j \le [(n+1)/2]} |S_j| \ge \epsilon n/c^2, |S_n| \le \delta)/P(|S_n| \le \delta) \\ &= \sum_{j=1}^{[(n+1)/2]} \lim_{\delta \downarrow 0} P(\tau = j, |S_n| \le \delta)/P(|S_n| \le \delta) \\ &= \sum_{j=1}^{[(n+1)/2]} \lim_{\delta \downarrow 0} \int P(\tau = j, S_j \in du), |u + \sum_{k=j+1}^{n} \xi_k| \le \delta)/P(|S_n| \le \delta) \\ &= \sum_{j=1}^{[(n+1)/2]} \lim_{\delta \downarrow 0} \int P(\tau = j, S_j \in du) P(|u + S_{n-j}| \le \delta)/P(|S_n| \le \delta) \end{aligned}$$

by the i.i.d. property of $\xi_1, \xi_2, \ldots, \xi_n$.

Since by assumption ξ_i has a bounded density A, for each n

$$S_n/(\sigma\sqrt{n}) \text{ has a bounded density.} \tag{2.49}$$

Call it h_n. By using a local limit theorem for densities in Petrov (1972; p. 198),

$$\lim_n\{\sup_n |h_n(s) - (exp - 2^{-1}s^2)/\sqrt{2\pi}|\} = 0. \tag{2.50}$$

It follows from (2.49) and (2.50) that

$$h_n(s) \leq M \ \forall n \geq 1, \ \forall s \in \boldsymbol{R}$$

holds for a certain constant M, and

$$\lim_n h_n(0) = 1/\sqrt{2\pi}.$$

For a fixed j and n with $j \leq [(n+1)/2]$ and n so large that $n \geq 4$ and $h_n(0) \geq 1/(2\sqrt{2\pi})$,

$$\begin{aligned}
&\lim_{\delta\downarrow 0} P(|u + S_{n-j}| \leq \delta)/P(|S_n| \leq \delta) \\
&= \lim_{\delta\downarrow 0}\left[\int_{(-\delta-u)/(\sigma\sqrt{n-j})}^{(\delta-u)/(\sigma\sqrt{n-j})} h_{n-j}(s)ds \Big/ \int_{-\delta/(\sigma\sqrt{n})}^{\delta/(\sigma\sqrt{n})} h_n(s)ds\right] \\
&= (\sqrt{n}/\sqrt{h-j})h_{n-j}(-u/(\sigma\sqrt{n-j}))/h_n(0) \\
&\leq 2M(2\sqrt{2\pi}) \leq 12M.
\end{aligned}$$

Hence by (2.48) and Lebesgue's bounded convergence theorem,

$$\begin{aligned}
&W_n(x : \|x\|_\infty > \epsilon(\sqrt{n}/c) \\
&\qquad \leq 2\sum_{j=1}^{[(n-1)/2]} \int 12MP(\tau = j, S_j \in du) \\
&\qquad = 24\sum_{j=1}^{[(n-1)/2]} P(\tau = j) \\
&\qquad = 24P\left(\max_{1\leq j\leq[(n+1)/2]} |S_j| > \epsilon n/c^2\right) = 0(n^{-1})
\end{aligned}$$

by (2.47). This proves (2.43).

Q.E.D.

This result tells us that if we are dealing with large configurations with strong couplings we can use Gaussian processes to approximate the prior.

This will lead to *dramatic speed up, both for pattern synthesis and analysis,* and will enable us to use large sweep areas.

Indeed, consider a Gaussian stochastic process $\{x_i;\ i = 0, 1, \ldots\ldots n-1\}$, real valued to begin with, which is Markovian in the sense of our cyclic connector graph σ. Without loss of generality we shall assume that the means are zero so that the density can be written as

$$f(x) \propto exp - \frac{1}{2} x' Q x \tag{2.51}$$

where, because of the Markovian assumption, Q is a tridiagonal (in the cyclic sense) matrix so that $q_{ij} = 0$ unless $i = j$ or $i = j \pm 1$ (recall that we use modular arithmetic for the i-values). The following representation will be useful.

Theorem 2.3. *The first-order Markov processes on the n-cyclic graph are given by densities in expression (2.4) with quadratic form*

$$\underline{x}' Q \underline{x} = \sum_{i=0}^{n-1} v_i (x_i - a_i x_{i+1})^2 \tag{2.52}$$

where $\{v_i\}$ and $\{a_i\}$ are such that Q is positive definite.

Proof: Without loss of generality we can assume, by rescaling, that the $m_{ii} = c_i$'s are equal to one and the $m_{ii+1} = b_i$'s are nonzero, for if a given b_i (or c_i) were zero, then the problem would reduce to that of a linear graph for which such a characterization (expression (2.52) above) is classical and given in Feller, Vol. II, p. 96. Equating the coefficients of the quadratic forms

$$1 = v_i + v_{i-1} a_{i-1}^2 \tag{2.53}$$

$$i = 0, 1, \ldots, n-1$$

$$b_i = -v_i a_i \tag{2.54}$$

we have

$$1 = v_i + v_{i-1} \left(\frac{b_{i-1}}{v_{i-1}} \right)^2 .$$

Letting $B_i = b_i^2$, we have the nonlinear difference equation

$$v_i = 1 - \frac{B_{i-1}}{v_{i-1}}, \quad i = 0, 1, \ldots, n-1 \tag{2.55}$$

and by n repeated substitutions we obtain for each i the following (cyclical) continued fraction:

$$v_i k = 1 - \cfrac{B_{i-1}}{1 - \cfrac{B_{i-2}}{\ddots \atop 1 - \cfrac{B_{i+1}}{1 - \cfrac{B_i}{v_i}}}} \qquad i = 0, 1, \ldots, n-1 \tag{2.56}$$

Defining $f_i(v)$ to be the R.H.S. of expression (2.55), it follows that

$$f_i(v) = \frac{\alpha_i v + \beta_i}{\gamma_i v + \delta_i} \tag{2.57}$$

is a linear fractional transformation constructed by composing n linear fractional transformations; the functions $f_i(\cdot)$ are defined on and take values in $\boldsymbol{R}_\infty$. One can obtain $\alpha_i, \beta_i, \gamma_i, \delta_i$ recursively:

$$\begin{aligned} f_i^{(0)}(v) &= 1 - \frac{B_i}{v} = \frac{\alpha_i^{(0)} v + \beta_i^{(0)}}{\gamma_i^{(0)} v + \delta_i^{(0)}} \\ f_i^{(k)}(v) &= 1 - \frac{B_{i+k}}{f_i^{(k-1)}(v)} = \frac{\alpha_i^{(k)} v + \beta_i^{(k)}}{\gamma_i^{(k)} v + \delta_i^{(k)}}, \quad k = 1, 2, \ldots, n-1 \end{aligned} \tag{2.58}$$

where

$$\begin{aligned} \alpha_i^{(k)} &= \alpha_i^{(k-1)} - B_{i+k}\gamma_i^{(k-1)} \qquad i, k = 1, 2, \ldots, n-1 \\ \beta_i^{(k)} &= \beta_i^{(k-1)} - B_{i+k}\delta_i^{(k-1)} \\ \gamma_i^{(k)} &= \alpha_i^{(k-1)} \\ \delta_i^{(k)} &= \beta_i^{(k-1)} \end{aligned} \tag{2.59}$$

and which, for each i, reduce to the following second-order linear difference equation with two sets of initial conditions:

$$\begin{aligned} \alpha_i^{(k)} &= \alpha_i^{(k-1)} - B_{i+k}\alpha_i^{(k-2)} \qquad k = 2, \ldots, n-1 \\ \alpha_i^{(0} &= 1, \ \alpha_i^{(1)} = 1 - B_{i+1} \end{aligned} \tag{2.60}$$

and

$$\begin{aligned} \beta_i^{(k)} &= \beta_i^{(k-1)} - B_{i+k}\beta_i^{(k-2)} \\ \beta_i^{(0} &= \beta_i^{(1)} = -B_i \end{aligned} \tag{2.61}$$

The coefficients of expression (2.57) are obtainable as $\alpha_i = \alpha_i^{(n-1)}, \beta_i = \beta_i^{(n-1)}, \gamma_i = \gamma_i^{(n-1)}$, and $\delta_i = \delta_i^{(n-1)}$. An alternative formulation of (2.59), using the homomorphism from Möbius transformations (from $\boldsymbol{R}_\infty$ to $\boldsymbol{R}_\infty$) onto $GL(2; \boldsymbol{R})$, is the following matrix product:

(2.62)

$$\begin{pmatrix} \alpha_i & \beta_i \\ \gamma_i & \delta_i \end{pmatrix} = \begin{pmatrix} 1 & (-B_{i-1}) \\ 1 & 0 \end{pmatrix} \times \begin{pmatrix} 1 & (-B_{i-2}) \\ 1 & 0 \end{pmatrix} \times \cdots \times \begin{pmatrix} 1 & (-B_i) \\ 1 & 0 \end{pmatrix}$$

Since by assumption M is positive definite, it follows that each matrix on the R.H.S. of expression (2.62) has positive determinant and we thus have:

$$\alpha_i \delta_i - \gamma_i \beta_i > 0.$$

From expression (2.57) one observes that each $f_i(\cdot)$ has a pole at $(-\delta_i/\gamma_i)$; we wish to show that each $f_i(\cdot)$ has at least one positive fixed point and that zero and ∞ are not fixed points of f_i. Consider the following tridiagonal matrices:

$$T_i = \begin{pmatrix} 1 & b_{i+1} & 0 & \dots & 0 \\ b_{i+1} & 1 & b_{i+2} & \dots & 0 \\ 0 & b_{i+2} & 1 & & \\ \vdots & & \ddots & & \\ & & & b_{i-2} & 0 \\ & & b_{i-2} & 1 & b_{i-1} \\ 0 & \dots & 0 & b_{i-1} & 1 \end{pmatrix} \quad i = 0, 1, \dots, n-1$$

The matrices $\{T_i\}$ are positive definite since they are principal minors of M. Consider the determinants of the principal minors of increasing order, k, of T_i, starting with the $(1,1)$ element, its order being defined as zero. Because of the tridiagonal nature of T_i, these determinants satisfy the above second-order difference equation with initial conditions given by expression (2.60). Therefore, by uniqueness we have that the $\alpha_i^{(k)}$, $k = 0, 1, \dots, n-1$ are given as the positive determinants for the principal minors of T_i and thus that

$$f_i(\infty) \stackrel{\text{def}}{=} \frac{\alpha_i}{\gamma_i} = \frac{\alpha_i^{(n-1)}}{\alpha_i^{(n-2)}}$$

is finite. If we replace the $(1,1)$ element of T_i with $(-B_i)$ and $(1,2)$ and $(2,1)$ elements with zero, the determinants of the principal minors of increasing order will now all be strictly negative and satisfy the second-order difference equation with initial conditions given by expression (2.61), thus determining the $\beta_i^{(k)}$, $k = 0, 1, \dots, n-1$, from which it follows that

$$f_i(0) \stackrel{\text{def}}{=} \frac{\beta_i}{\delta_i} = \frac{\beta^{(n-1)}}{\beta^{(n-2)}}$$

is positive. We thus also know that α_i and γ_i are positive, β_i and δ_i are negative, and $\alpha_i\delta_i - \Gamma_i\beta_i > 0$. We have shown that zero and ∞ are not fixed points for $f_i(\cdot)$, $i = 0, 1, \ldots, n-1$; we need to show that each $f_i(\cdot)$ has at least one positive fixed point. By expression (2.57) it follows that one needs to show that the discriminant:

$$(\delta_i - \alpha_i)^2 + 4\beta_i\gamma_i$$

is nonnegative for the quadratic equation

$$\gamma_i v^2 + (\delta_i - \alpha_i)v - \beta_i = 0$$

since we know that $-(\delta_i - \alpha_i)$ is positive. Consider the following two matrices (A_1 and A_2) and their respective spectral decompositions:

$$\begin{pmatrix} \alpha_i & \gamma_i^{1/2}|\beta_i|^{1/2} \\ \gamma_i|\beta_i|^{1/2} & -\delta_i \end{pmatrix} = P_1'\Lambda_1 P_1,$$

and

$$\begin{pmatrix} -\delta_i & \gamma_i^{1/2}|\beta_i|^{1/2} \\ \gamma_i^{1/2}|\beta_i|^{1/2} & \alpha_i \end{pmatrix} = P_2'\Lambda_2 P_2$$

They have the same eigenvalues (one positive, one negative), their (common) trace, $\alpha_i - \delta_i$, is positive, and their corresponding eigenvectors differ by a reversal of coordinates:

$$\Lambda_1 = \Lambda_2 = \begin{pmatrix} \lambda_1 & 0 \\ 0 & \lambda_2 \end{pmatrix}, \; P_2 = \begin{pmatrix} 0 & 1 \\ 1 & 0 \end{pmatrix} P_1$$

We thus have that the sum of A_1 and A_2 has the following spectral decomposition:

$$\begin{pmatrix} (\alpha_i - \delta_i) & 2\Gamma_i^{1/2}|\beta_i|^{1/2} \\ 2\Gamma_i^{1/2}|\beta_i|^{1/2} & (\alpha_i - \delta_i) \end{pmatrix} = P_1' \begin{pmatrix} \lambda_1 + \lambda_2 & 0 \\ 0 & \lambda_1 + \lambda_2 \end{pmatrix} P_1$$

and since $\lambda_1 + \lambda_2 = \alpha_i - \delta_i$ is positive, it follows that $A_1 + A_2$ is positive definite and its determinant, the desired discriminant, is positive.

Therefore, for each i, $f_i(\cdot)$ has at least one positive fixed point and neither zero nor ∞ are fixed points. Fix i and let v_i be a positive fixed point of $f_i(\cdot)$ and define v_{i+1} by expression (2.55):

$$v_{i+1} \stackrel{\text{def}}{=} 1 - \frac{B_i}{v_i}$$

By expression (2.56) for $i+1$, it follows that v_{i+1} is a fixed point of $f_{i+1}(\cdot)$, for which we know that it is neither zero nor ∞. Continuing in this manner, v_i, $i = 0, 1, \ldots, n-1$, are defined; using expression (2.54) the a_i's are determined. Nothing so far prevents some of the v_i's from being negative; however, by the equality of the two quadratic forms in (2.51) and (2.52), it follows that Q must be positive definite since expression (2.52) gives the quadratic form as a linear combination of squares of linear functions.

Q.E.D.

But then we can change variables by the non-singular transformation

$$e_i = x_i - a_i x_{i-1};\ i = 0, 1, \dots n-1 \tag{2.63}$$

We are later going to take x as an entry in a matrix representation of the similarity group elements. Note that the $\{e_i\}$ are stochastically independent and play the role of innovations on our cyclic graph σ. Solving successively for the x's we get

$$\begin{aligned} x_i &= e_i + a_i x_{i-1} \\ &= e_i + a_i e_{i-1} + a_i a_{i-1} x_{i-2} \\ &\dots \\ &= e_1 + a_i e_{i-1} + a_i a_{i-1} e_{i-2} + \cdots + a_i a_{i-1} \dots a_{i+1} x_i \end{aligned}$$

But we know that $a_0 a_1 \dots a_{n-1} \neq 1$, guaranteeing non-singularity of our quadratic form, so that we can solve for x_i which gives the useful representation

$$x_i = \frac{1}{a_0 a_1 \dots a_{n-1} - 1}[e_i + a_i e_{i-1} + a_i a_{i-1} e_{i-2} + \cdots + a_i a_{i-1} \dots a_{i+1} e_{i+1}]$$

Since independent Gaussian stochastic variables can be simulated by fast algorithms we are able to generate many x-values at once (even the whole x-process) directly so that we can use large sweep areas. The above will be modified in Sections 4,5 to incorporate regularity constraints.

We can rewrite the quadratic form

$$x^T M x = \sum_{i=0}^{n-1} v_i (x_i - a_i x_{i-1})^2 = \Sigma \alpha_i x_i^2 + \Sigma \beta_i (x_i - x_{i-1})^2$$

where

$$\begin{cases} \alpha_i = v_i + v_{i+1} a_{i-1}^2 - a_i v_i - a_{i+1} v_{i+1} \\ \beta_i = a_i v_i \end{cases} \tag{2.64}$$

and we identify, putting $A = exp - \frac{a}{2}$, $Q = exp - \frac{q}{2}$

$$\begin{cases} q_i(x_i) = \alpha_i x_i^2 \\ a_i(x_i, x_{i-1}) = \beta_i (x_i - x_{i-1})^2 \end{cases}$$

Note that the sign of β equals that of the coupling coefficient a_i.

In the homogeneous case (2.64) reduces to

$$\begin{cases} \alpha = v(1-a)^2 \\ \beta = av \end{cases}$$

Making the value of α close to one will make α small, unless we make v large, so that we will be close to singularity, which will strongly influence the behavior of the picture ensemble that we will synthesize in Section 4.

We mention in passing that the stochastic difference equation (2.63) has a continuous analog, a stochastic differential equation in which the argument takes values in the circle, but this will not be explored here.

To see how the above is realized for the choice $S = GL(2)$, say that we update sites $i = i_{left}, \ldots i_{right}$ in s that leave the vector CHORD, see Figure 6,

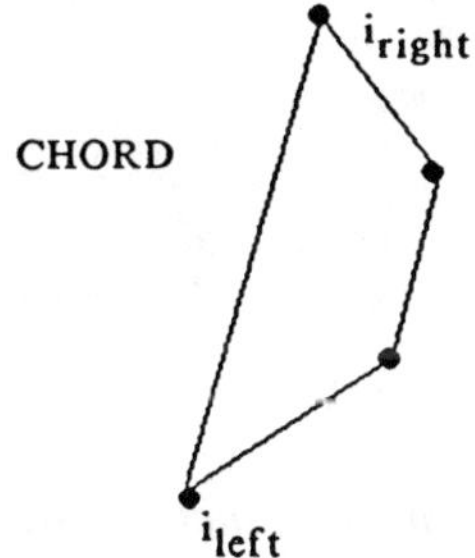

FIGURE 6

from the current vertex $V_{i_{left}}$ to $V_{i_{right}}$ invariant. If ACHORD is the homologous vector in the template we should have

$$S \text{ ACHORD} = \text{CHORD} \tag{2.65}$$

Let us start by considering the x-coordinate only of this vector equation.

Writing $s = e + t$, and with the notation

$$\begin{cases} t = \begin{pmatrix} t_{00} & t_{01} \\ t_{10} & t_{11} \end{pmatrix} \\ ACHORD \qquad = Col(a_0, a_1) \\ CHORD \qquad = Col(c_0, c_1) \end{cases}$$

the first component of (2.65) can be written

$$t_{00}a_0 + t_{01}a_1 = c_0 - a_0 = d_0 \tag{2.66}$$

With $t_{00} = x, t_{01} = y, d_0 = d$ we can write the quadratic form K in the exponent of the normal density (except for the factor $-1/2\sigma^2$ as

$$K = (x - m_0)^2 + (y - m_1)^2 = (x - m_0)^2 + \left(\frac{a_0}{a_1} + \frac{d}{a_1} + m_1\right)^2$$

where m_0 and m_1 are the conditional means of the t_{00} and t_{01} respectively in the Markov process conditioned by the values at i_{left} and i_{right}. Hence

$$K = x^2\left(1 + \frac{a_0^2}{a_1^2}\right) - 2x\left(m_0 + \frac{a_0}{a_1^2}d - \frac{a_0}{a_1}m_1\right) + \text{constant}.$$

This implies that x only will have the mean values

$$\begin{cases} m_x = \frac{1}{A_2}(m_0 a_1^2 + a_0 d - a_0 a_1 m_1) \\ m_y = \frac{1}{A_2}(-m_0 a_0 a_1 + a_1 d + a_0^2 m_0) \\ A_2 = a_0^2 + a_1^2 \end{cases}$$

and the conditional variances

$$\begin{cases} \sigma_x^2 = \frac{a_1^2}{A_2}\sigma^2 \\ \sigma_y^2 = \frac{a_0^2}{A_2}\sigma^2 \end{cases}$$

If $u = N\left(0, \frac{\sigma^2}{A_2}\right)$, and recalling relation (2.66) we can simulate (x, y) simply

$$\begin{cases} x = a_1 u + (m_0 a_1^2 + a_0 d - a_0 a_1 m_1)/A_2 \\ y = -a_0 u + (-m_0 a_0 a_1 + a_1 d + m_1 a_0^2)/A_2 \end{cases}$$

This is implemented in the program RANDOMGROUP2 called by GLOBAL2. The latter will be used only for the situation that a single site i is updated, so that $i_{left} = i - 1, i_{right} = i + 1$, or for the whole chain $i = 1, 2, \ldots n - 1$, so that $i_{left} = 1, i_{right} = n \equiv 0$.

Now let us turn to the similarity group $S = 0(2)\times US(2)$. The simulation in this case is done by RANDOMGROUP3.

RANDOMGROUP3, called by GLOBAL3, allows for the various template arcs, lying between $\ell = i_{left}$ and $r = i_{right}$, to each have a different matrix group element. For the case of $0(2)\times US(2)$, each matrix is identified by its first row:

$$S^{(i)} = \begin{pmatrix} 1 + t_{00}^{(i)} & t_{01}^{(i)} \\ -t_{01}^{(i)} & 1 + t_{00}^{(i)} \end{pmatrix} = T^{(i)} + I_2.$$

Consequently, if we are to update the portion of the configuration from ℓ to r, we need the conditional distribution of

$$\underline{t} = ((t_{00}^{(i)}, t_{01}^{(i)}),\ i = \ell, \ldots, r - 1) \text{ given } (t_{00}^{(\ell-1)}, t_{01}^{(\ell-1)}, t_{00}^{(r)}, t_{01}^{(r)});$$

in addition, one needs to further condition on the configuration remaining regular. Since the two processess $\{t_{00}\}$ and $\{t_{01}\}$ are independent, first-order Gauss Markov processes on the n-cyclic graph, they remain independent when we condition on their values at $(\ell-1)\ell$ and r. From Section 2.2 it follows that the conditional distribution of $\{t_{00}^{(i)},\ i=\ell,\ldots,r-1\}$ given $(t_{00}^{(\ell-1)}, t_{00}^{(r)})$ has a covariance matrix whose inverse is:

$$\begin{pmatrix} \left(\frac{1}{\sigma_\ell^2}+\frac{a_{\ell+1}}{\sigma_{\ell+1}^2}\right) & \left(\frac{-a_{\ell_1}}{\sigma_{\ell+1}^2}\right) & & & 0 \\ \left(\frac{-a_{\ell+1}}{\sigma_{\ell+1}^2}\right) & \left(\frac{1}{\sigma_{\ell+1}^2}+\frac{a_{\ell+2}}{\sigma_{\ell+2}^2}\right) & \left(\frac{-a_{\ell+2}}{\sigma_{\ell+2}^2}\right) & 0 & \vdots \\ 0 & \ddots & \ddots & & 0 \\ \vdots & & & & \left(\frac{-a_{r+1}}{\sigma_{r-1}^2}\right) \\ 0 & & \left(\frac{-a_{r+1}}{\sigma_{r-1}^2}\right) & & \left(\frac{1}{\sigma_{r+1}^2}+\frac{a_r}{\sigma_r^2}\right) \end{pmatrix}$$

and whose mean vector is:

$$\left(\frac{a_\ell}{\sigma_\ell^2}t_{00}^{\ell-1}, 0, 0, \ldots, 0, \frac{a_r}{\sigma_r^2}t_{00}^{(r)}\right)$$

where the $\underline{a}$ and $\underline{\sigma}^2$ vectors are the parameters (ABOND and SIG2 in the code) for the $\{t_{00}\}$ process. Similarly one obtains the analogous expression for the $\{t_{01}\}$ process and because of independence one obtains the desired covariance matrix, Σ and mean vector, $\underline{\mu}$, for the joint variables, $\underline{t}$. To insure regularity, the t's are constrained to satisfy:

$$\sum_{i=\ell}^{r-1} T^{(i)} g_i^{(0)} = (\text{CHORD-ACHORD})$$

where

$$\sum_{i=\ell}^{r-1} g_i^{(0)} = \text{ACHORD}$$

and which naturally present themselves as two linear constraints, written as a $(2(r-\ell),2)$ matrix A, on the vector of matrix components, $\underline{t}$. The concern now is the simulation from the singular Gaussian measure for:

$$\underline{t}|(A'\underline{t} = (\text{CHORD-ACHORD}))$$

whose covariance matrix is

$$\bar{\Sigma} - \bar{\Sigma}A(A'\bar{\Sigma}A)^{-1}A'\bar{\Sigma} \tag{2.67}$$

and mean vector is

$$\underline{\mu} + \bar{\Sigma}A(A'\bar{\Sigma}A)((\text{CHORD-ACHORD}) - A'\underline{\mu}). \tag{2.68}$$

To calculate a "square root" of the singular covariance matrix we first calculated L, the lower triangular matrix from the Cholesky Decomposition of $\bar{\Sigma}$; rewriting expression (2.67) as:

$$L[I - L'A(A'\Sigma A)^{-1}A'L]L' \tag{2.69}$$

and by calculating the bracketed expression, P, a symmetric, idempotent matrix, one can then use the matrix product LP as a "square root". RANDOMGROUP3 generates SS such realizations simultaneously and outputs SS realizations of the $(r - \ell)$ operators, $S^{(i)}$, sitting on the edges of the template from vertices ℓ to r. An analogous program, RANDOMGROUP1, called by GLOBAL1, generates realizations from $GL(2; \boldsymbol{R})$.

3
Pattern Analysis

The task of restoring a noisy image is facilitated by having models and algorithms already available for synthesis. When we developed the code it was both possible and convenient to use the same code for both purposes with a switch called, SYNTHESIS, to be turned on or off as required. Other switches will be used later. But first we have to clear up an issue concerning the simulation of the posterior (1.12).

§3.1. This density factors into the prior $p(c)$ and a second factor that represents the stochastic deformation $\mathcal{D} : C \to I^{\mathcal{D}}$. We have seen how we can simulate $p(c)$ fast but the second factor has little analytical structure that can be exploited. To handle this difficulty let us look at the following more general situation.

In a finite dimensional sample space $X = \boldsymbol{R}^d$ consider an absolutely continuous probability measure given by a density

$$f(x) = f_1(x) f_2(x); \qquad x \in X \tag{3.1}$$

where f_1 is a density which we know how to simulate. If f itself is not possible to simulate directly we can use the "*factored sampling*" scheme from the following result.

Theorem 3.1. *Let $x_1, x_2, \ldots x_N$ be an i.i.d. sample from f_1 and choose one of them at random, say $x_i = y$, with probabilities*

$$p_j = \frac{f_2(x_j)}{\sum_1^N f_2(x_j)}; \; j = 1, 2, \ldots N. \tag{3.2}$$

Then the probability distribution of y tends weakly as $N \to \infty$ to that of f.

Proof: The denominator on the right-hand side of (3.2) is positive with probability one so that the probabilities p_j are well defined. The density of y, call it g_N, is then

(3.3)

$$g_N(y) = N \underbrace{\int \cdots \int}_{N-1} \frac{f_1(y)f_2(y)}{\sum_1^{n-1} f_2(x_j) + f_2(y)} f_1(x_1)\ldots f_1(x_{N-1})dx_1\ldots dx_N$$

which can be written as $f(y)J_N(y)$ with the expected value (in terms of the f_1 density)

$$J_N(y) = E_1\left[\frac{1}{\frac{1}{N}\sum_1^{N-1} F_j + f_2(y)}\right] \tag{3.4}$$

with the non-negative random variables $F_j = f_2(x_j)$ and where the value of $f_2(y)$ is positive a.s. According to the law of large numbers

$$\begin{aligned}\frac{1}{N}\sum_1^{N-1} F_j \to E_1(F_1) &= \int_{-\infty}^{\infty} f_2(x)f_1(x)dx \\ &= \int_{-\infty}^{\infty} f(x)dx = 1 \ a.s.\end{aligned} \tag{3.5}$$

Hence for a.e. y we have pointwise convergence

$$\lim_{N\to\infty} g_N(y) = f(y) \tag{3.6}$$

Since the limiting function has integral one the statement in the theorem follows from Scheffe's theorem (see e.g. Billingsley, 1968. p. 223).

Q.E.D.

We shall interpret f_1 as the prior density p, which we know how to simulate fast, and let $f_2 = p(I^{\mathcal{D}}|I)$. Then generate a sample $(I_1, I_2, \ldots I_N)$ from p_1, compute the values of $p(I^{\mathcal{D}}|I_j)$; $j = 1, 2, \ldots N$, and select one of the I_j at random with probabilities proportional to the $p(I^{\mathcal{D}}|I_j)$.

This method, which may remind the reader of the acceptance/rejection technique for Monte Carlo simulation, (see e.g. Hammersley-Handscomb (1964), p. 36), which is, however, not applicable here, because the density is too peaked, has been used extensively in the computer experiment to be described below. We know little about the speed of convergence as $N \to \infty$; it may be possible to determine it by applying some version of the central limit theorem, instead of the law of large numbers, in the proof.

In our restoration experiments we shall, at first, assume a simple and not very realistic deformation mechanism:

(3.7)

$\mathcal{D} : I^{\mathcal{D}}(z) = N(m_\nu, \tau^2)$ with $m_\nu = m_{in}$ if $z \in I$ and $m = m_{out}$ if $z \notin I$

together with stochastic independence between the $I^{\mathcal{D}}$-values. Hence we have three nuisance parameters m_{in}, m_{out}, τ^2 that will be estimated from data since they will vary with light conditions and depend upon visual noise.

In Plate 3 we show the histogram of grey level values for a clear picture. It shows two well separated peaks which could probably be determined well by estimating m_{in}, m_{out}, τ^2 from a mixture model with two Gaussian densities of the same variance. This is essentially a deconvolution procedure.

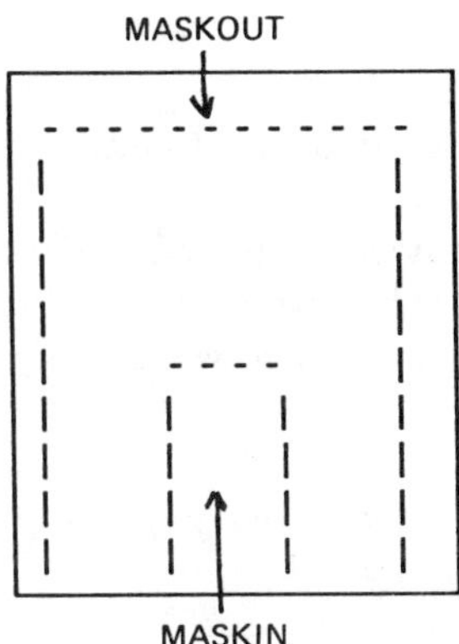

FIGURE 7

But this is not typical for noisy pictures. For one picture with moderate S/N ratio the histogram was displayed in Plate 2. Here it is doubtful whether the peaks could be determined by the same sort of procedure.

We have, however, additional knowledge, as will often be the case for bio/medical imaging, in that we possess strong prior information of how hands look and where in the picture they can be expected to be placed. In Figure 7 we indicate two masks, MASKIN that is expected to be mainly inside the hand, and another, MASKOUT, that will mainly be outside of the hand set.

We can then use the straightforward estimates

(3.8)
$$\begin{cases} m_{in}^* &= \frac{1}{n_{in}} \sum_{z \in MASKIN} I^{\mathcal{D}}(z) \\ m_{out}^* &= \frac{1}{n_{out}} \sum_{z \in MASKOUT} I^{\mathcal{D}}(z) \\ (\tau^*)^2 &= \frac{1}{n} \left\{ \sum_{z \in MASKIN} [I^{\mathcal{D}}(z) - m_{in}^*]^2 + \sum_{z \in MASKOUT} [I^{\mathcal{D}}(z) - m_{out}^*]^2 \right\} \end{cases}$$

and n_{in}, n_{out} are the number of lattice points in respective masks.

The second factor in the posterior (1.2) then takes the simple form, with $I = Rc$,

(3.9)

$$exp-\frac{1}{2(\tau^*)^2}\left\{\sum_{z\in I}[I^{\mathcal{D}}(z)-m^*_{in}]+\sum_{z\notin I}[I^{\mathcal{D}}(z)-m^*_{out}]^2\right\}=exp\left(-\frac{1}{2(\tau^*)^2}K\right)$$

Note that we need only calculate two sums. Indeed, we can write the quadratic form

$$(3.10)\quad K = \text{constant } -2(m^*_{in}-m^*_{out})\sum_{z\in I} I^{\mathcal{D}}(z)+\left[(m^*_{in})^2-(m^*_{out})^2\right]\sum_{z\in I} 1$$

where the constant depends upon $I^{\mathcal{D}}$ (which should be treated as fixed) but not upon I.

In Figure 8 we show part of the boundary and consider the sweep area from $i = i_{left}$ to i_{right}. When we consider a sample of N hypothetical boundaries from $i = i_{left}$ to i_{right} to which to apply Theorem 3.1 we can replace (3.10) by

$$(3.11)\qquad K = \text{constant } -2(m^*_{in}-m^*_{out})S_x+\left[(m^*_{in})^2-(m^*_{out})^2\right]S_1$$

where

$$(3.12)\qquad \begin{cases} S_x = \sum_{z\in F} I^{\mathcal{D}}(z) \\ S_1 = \sum_{z\in F} 1 \end{cases}$$

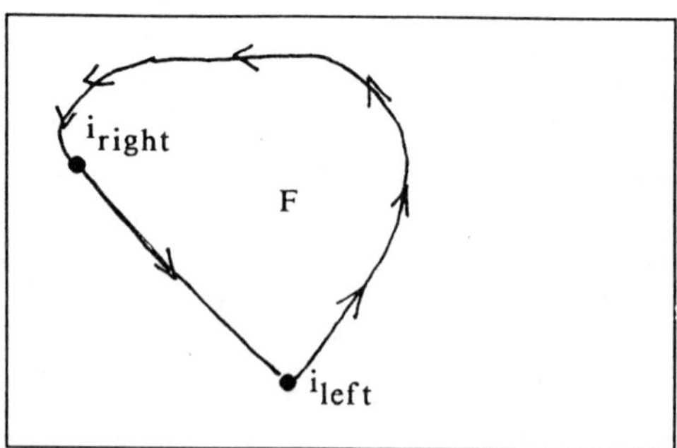

FIGURE 8

and F is the set bounded by the boundary from i_{left} to i_{right} and the straight chord back to i_{left}. This is correct if we evaluate (3.12) interpreted as sums over oriented sets. For example if the chord cuts the boundary once we will have one positively oriented and another negatively oriented set.

These "area integrals" over areas bounded by contours Γ will be evaluated in the program DEFPROB as "contour integrals"

$$\oint_{\Gamma} y dx \tag{3.13}$$

appropriately discretized. This leads to a fairly fast algorithm but can probably be speeded up by an order of magnitude.

Often these sets are fairly small, depending upon the size of the sweep areas. In Figure 8 we shall choose the rectangle containing the sweep area generators as small as possible so that it contains all N realizations of the hypothetical boundaries but nothing more.

§3.2. In some of our experiments this worked well but not in others. The deformation model (3.7) is not realistic *if the background lighting is strongly non-uniform.* It is well known that this can happen even when the picture appears to have uniform background. As an example consider Plate 4, which is quite a noisy hand picture, and the corresponding profile and histogram in Plate 5. The profile indicates a strong gradient in the background lighting which must be taken into account when designing the deformation model. This becomes even clearer when one examines the histogram: instead of having two well separated peaks only one can be discussed. For example, see Plates 4 and 5 (further details given in Section 5.3.1).

To deal with this difficulty we shall replace (3.7) by the assumption that the means $m_{in}(z), m_{out}(z)$ are allowed to vary, but only slowly over the picture. We shall carry out *local estimation* of the nuisance parameters as follows.

Instead of the global estimates (3.8) let us estimate m_{in}, m_{out}, τ considered as a constant over the small rectangle Q in Figure 8, and do this by maximum likelihood.

For any of the N hypothetical boundaries we should minimize K/τ^2 with respect to m_{in}, m_{out}, and τ. This gives us a likelihood proportional to

$$\left(\frac{1}{\tau^*}\right)^{n_{total}} \tag{3.14}$$

with the obvious estimators

$$\begin{cases} m^*_{in} & = \frac{1}{n_{in}} \sum\limits_{z\in I} I^{\mathcal{D}}(z) \\ m^*_{out} & = \frac{1}{n_{out}} \sum\limits_{z\notin I} I^{\mathcal{D}}(z) \\ (\tau^*)^2 & = \frac{1}{n_{in}n_{out}} \left\{ \sum\limits_{z\in I} [I^{\mathcal{D}}(z) - m^*_{in}]^2 + \sum\limits_{z\notin I} [I^{\mathcal{D}}(z) - m^*_{out}]^2 \right\} \end{cases} \tag{3.15}$$

where n_{in}, n_{out} are the number of lattice points inside and outside $I \cap Q$.

The estimates (3.14) can be calculated by evaluating six sums, namely

$$(3.16) \qquad \begin{cases} S_1 &= \sum_{z \in I \cap Q} 1 \\ S_x &= \sum_{z \in I \cap Q} I^{\mathcal{D}}(z) \\ S_{x2} &= \sum_{z \in I \cap Q} I^{\mathcal{D}}(z) \end{cases}$$

and the corresponding ones over $I^c \cap Q$. This computation will appear in the code.

Another modification of the deformation mechanism $\mathcal{D}$ is to allow *dependent noise.* If we only wish to express what order of magnitude the correlation length should have it may suffice to use a Markov model on the $LV \times LH$ lattice, say with a density $exp - \frac{1}{2}K$ with the quadratic form

$$(3.17) \qquad \sum_{x,y} K(x,y;x',y')[I^{\mathcal{D}}(x,y) - m(x,y)][I^{\mathcal{D}}(x',y') - m(x',y')]$$

with

$$(3.18) \qquad m(x,y) = \begin{cases} m_{in} \ if \ (x,y) \in I \\ m_{out} \ if \ (x,y) \notin I \end{cases}$$

and

$$(3.19) \qquad K(x,y;x',y') = \begin{cases} k_0 \ if \ (x,y) = (x',y') \\ k_1 \ if \ (x,y) \text{ neighbors } (x',y') \\ 0 \text{ else} \end{cases}$$

This $\mathcal{D}$ has not been implemented in code.

Another $\mathcal{D}$ is when the background is known a priori to be darker in some places than in others. If this is given by *non-uniform intensity* $\mu(x,y)$, and if we assume, somewhat questionably, that intensities add, then we can simply replace $I^{\mathcal{D}}(x,y)$ by $I^{\mathcal{D}}(x,y) - \mu(x,y)$ and do the analysis later on. This has not been tried.

The algorithms developed give us a powerful tool for handling the second task (ii) in section 1. Indeed if there is a pathology along the boundary from site i' to i'', this will show up in the corresponding values for the group-valued stochastic process s_i.

By examining the norms $n_i = \|s_i - e\|$, as induced by $(s_{00}, s_{01}) \in \boldsymbol{R}^2$, the pathology will tend to make n_i large for $i \in (i', i'')$. Thus a *pathology detector* can be constructed by observing for what i-values n_i/ν_i takes on values larger than some threshold.

4
Pattern Synthesis: Experiments

We shall now apply the random shape model to hand pictures and start with simulation experiments. This will help us to get an intuitive understanding of the model and the prior measure it represents, what is the influence of the parameters and what sort of picture ensembles it will generate.

Looking ahead to dealing with other bio/medical images we stress our research strategy: to *exploit the anatomical knowledge available*, both about typical shape and variation around it.

§4.1. We easily recognize the shape of a right human hand in spite of all the variation in its shape. There is variation from individual to individual, especially in size but also in proportion. But there is also more fine scale variation, say in the thickness of the fingers, the lengths between the joints. Then, for a given individual, the shape varies from time to time when he puts down his hand on the table: the angles of the fingers will vary as well as the flatness of the hand.

This has motivated us to introduce the following *multistage model*. Given the template c_{temp} in (1.9) we shall assume that the configuration c is given as

(4.1)

$$c = \sigma((s_0^{p-1} \dots s_0^1 s_0^0)g_0, \dots (s_i^{p-1} \dots s_i^1 s_i^0)g_i, \dots (s_{n-1}^{p-1} \dots s_{n-1}^1 s_{n-1}^0)g_{n-1})$$

where $s_i^\nu \in S^\nu$ and $S^0, \dots S^{p-1}$ are p similarity groups. Here p is the number of stages.

In the following experiments we have used several versions of (4.1), the most commonly used being

$$\text{(4.2)} \qquad \begin{cases} p &= 2 \\ S^0 &= GL(2) \\ S^1 &= 0(2) \times US(2) \end{cases}$$

and

$$(4.3) \qquad \begin{cases} p &= 3 \\ S^0 &= GL(2) \\ S^1 &= S^2 = 0(2) \times US(2) \end{cases}$$

with s_i^0 being constant, not varying with i. The other $(s_i^\nu; i = 0, 1, \ldots n-1)$ are piecewise constant over stretches of sites, with Markovian dependence between the values.

The stage $\nu = 0$ is meant to account for large scale variation between individuals and for orientation, the later stages represent fine structures of the shape, changing angles between fingers, changes in lengths, etc.

On the νth stage let us consider the stretches of sites over which the group valued process is constant $(i_0, i_1), (i_1, i_2), \ldots (i_{\mu-1}, i_0)$ with all arithmetic over site numbers understood to be modulo n. The sites $i_0(= 0), i_1, i_2, \ldots i_{\mu-1}$ are referred to as the *hinges.* For stage 0 we always choose $\mu = 1$, so that s_i^0 is not varying, just one randomly selected group element.

We have operated with $n = 256$ sites in the connector graph. For the higher stages we have used $\mu = 36$ or 46. Since $0(2) \times US(2)$ is a two-dimensional group and $GL(2)$ is four-dimensional we operate with $72 - 96$ degrees of freedom which reflects the statement in section 1 that we are dealing with high dimensional parameter spaces.

The idea of stages is reminiscent of the multi-scale, multi-grid methods in theoretical physics and numerical analysis which are intended to speed up convergence; see e.g. (Paddon-Holstein (1985)). The main reason why we have used it is, however, for another reason, namely the belief that *it mirrors essential properties of these biological patterns and of the way they are generated.*

§4.2. The hand pictures were acquired as follows. The subjects, all adult males, were asked to put their right hand on top of a rectangle above a light table. They should put the left side of the wrist at a specified mark on the rectangle and such that the hand was entirely within the rectangle. They were free to spread the fingers more or less and choose the general orientation; no other instructions were given.

With the light table lit the picture was captured using a digital camera, COHU Solid State camera, and using the image processing software package IMAGEPRO they were stored on the hard disk of a SPERRY AT. The resolution of the system was 512×480 but the pictures actually used were in a format $LV \times LH$, usually $LV = 128, LH = 120$, occasionally with even lower resolution. The number of grey level was usually 256.

The pictures were displayed on an ELECTROHOME monitor and photographs were taken of what appeared on the screen with a 35mm camera. This procedure was time consuming but was used since we did not have any color printer available originally.

In the analysis part of the experiments many sorts of visual noise were introduced, but for the synthesis we tried to obtain clear pictures undisturbed by noise. This takes considerable care, and even a picture that appears clean on the monitor will often show a good deal of noise when analyzed numerically, say thresholded or profiled.

To avoid reflections and other disturbances or distortions these pictures had to be captured at night with the room as dark as possible. Since this is a study in pattern theory, not in computer vision, we would have preferred not to have to worry about such issues, but they could not be avoided entirely.

§4.3. The software was written in modular fashion and with logical switches that could be turned on and off for different purposes. The Boolean variable SYNTHESIS, that has been mentioned, should be on for simulation, off for analysis. The variable LARGE should be on if the picture is so large that the high dimensional arrays created by the programs do not fit into memory. The value for RANDOM should always be on for simulation, and sometimes for analysis as will be discussed in Section 5.

We choose APL as the main programming language for the experiments. Some of the code for image handling of standard type was written in C. It had been questioned whether a mathematical computer experiment of this size could be done in APL, but this turned out to be possible. Indeed, it is difficult to see how such sophisticated software of unconventional and untried form could have been developed and modified in such a short time using any other programming language.

Most of the computations were done on our mainframe IBM 3090 in time sharing mode. Some of the standard image processing was done in C on the SPERRY.

The basic program modules are called GLOBAL2, GLOBAL3, etc; see Appendix A for program listings. GLOBAL2 implements the construction in Section 2.2. Executing ILEFT GLOBAL2 IRIGHT will produce a configuration with a constant $s \in GL(2)$ applied to all the sites from ILEFT to IRIGHT. The s-value is generated by RANDOMGROUP2.

For example 1 GLOBAL2 0 simulates the whole configuration except the segment between sites 0 and 1 by a rigid transformation from GL(2). Similarly 1 GLOBAL3 0, starting from c_{temp} as a circle (actually a regular 256-gon) gives us the deformed circle ensemble in Plate 6; here the s_i-process varies over the sites.

§4.4. The template was determined using the estimator in (1.13), where the 16 landmarks shown in Plate 7 were determined subjectively for each hand boundary c_ν. The boundaries were shown on the monitor and we decided interactively where to place the landmarks. The c_ν were obtained by first thresholding the clean picture and then computing the boundary by a simple edge detector. The boundaries were stored chain coded.

Between two consecutive landmarks 8 sites were placed equidistantly along the boundary. This procedure for estimating the template is clearly too ad hoc, but was used since, at the time, we had no objective method available. Such a method has since been developed and has been reported in Knoerr (1988).

In Plate 7 we show some boundaries (green curves) together with the template estimate (red curve) where the red dots indicate landmarks.

Let us apply GLOBAL2 to the template using 36 hinges. Three realizations from the resulting random ensemble are shown in the form of the corresponding 36-gons in Plate 8. This prior is quite spread out – the polygons cover most of the $LV \times LH$ rectangle and seems a reasonable choice for the following experiments.

Let us also synthesize hand shapes by calling HANDSYNTH3 which first executes GLOBAL2, makes the resulting shape a new template, and then applies GLOBAL3 with 36 hinges to it. We show the shape ensembles in Plate 9, all with SIGMAS equal to .3 and with varying SIG2 (corresponding to $1/v_i$) and ABOND (corresponding to a_i in formula (2.52)).

This and many other synthesis experiments have convinced us that this type of prior measure is what we need; the values of the parameters have to be discussed however.

§4.5. To get a better feeling for how SIG2 and ABOND influence the priors we have used expressions (2.67) and (2.68) to compute the standard deviation of the $x-$ and $y-$components of the jth vertex among the hinges. The resulting graphs are shown in Appendix C and we shall only discuss a few here (Figures 9-11).

We also show the effect of heterogeneous SIG2 and ABOND vectors in Figure 10, where the x and y standard deviations are different and are shown as 'rings' and 'stars' respectively, and two examples of strong couplings in Figure 11. Somewhat surprisingly the two latter graphs show a hand shape emerging; this is due to the influence of the template on the induced prior measure.

These graphs were helpful for planning the experiment. Preliminary experiments had given us the impression that the entries in the ABOND vector must be kept small, say in the range 0 to .2 to prevent excessive variability, and this was indeed done in the beginning. The analytical reason is that as the ABOND values increase, for example all equal and tending to one, the covariance operator tends toward singularity and the measure will eventually blow up.

The graphs above modified our view. It appears that even large ABOND values, say .6 to .8, keep the variability moderate for adequate choices of SIG1,SIG2. We therefore increased them drastically during the later stage of the experiment.

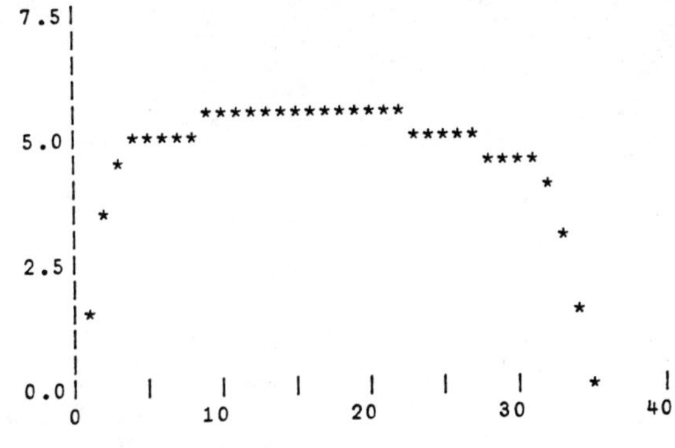

ABOND CONSTANT = 0.2
SIG2 CONSTANT = 0.01

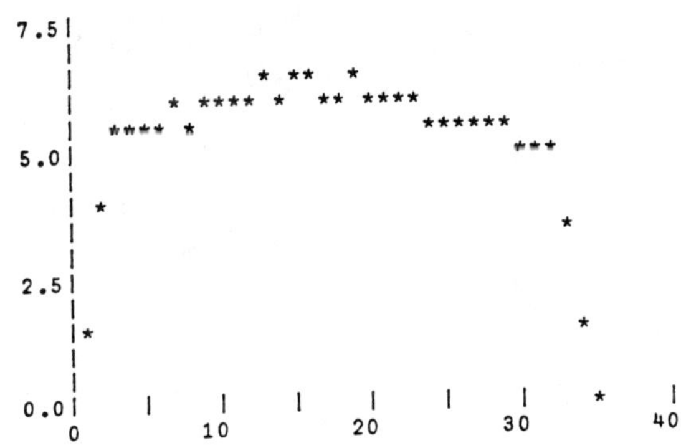

ABOND CONSTANT = 0.4
SIG2 CONSTANT = 0.01

FIGURE 9

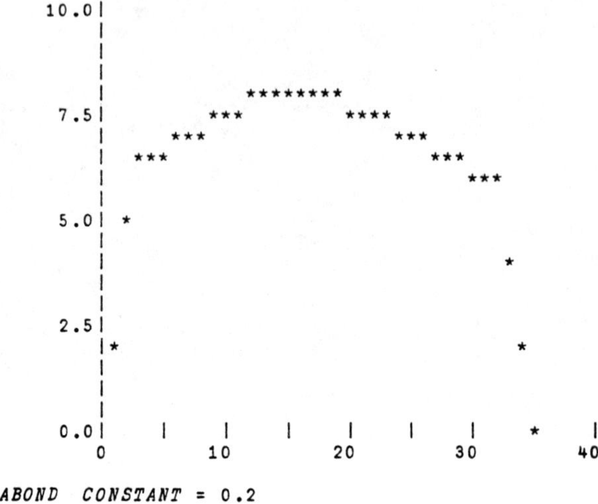

ABOND CONSTANT = 0.2
SIG2 CONSTANT = 0.02

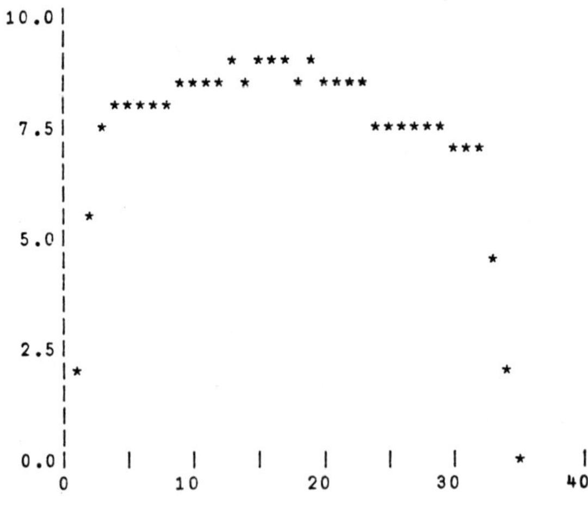

ABOND CONSTANT = 0.4
SIG2 CONSTANT = 0.02

FIGURE 9 (*continued*)

```
>SIG2←2 36ρ.02
```

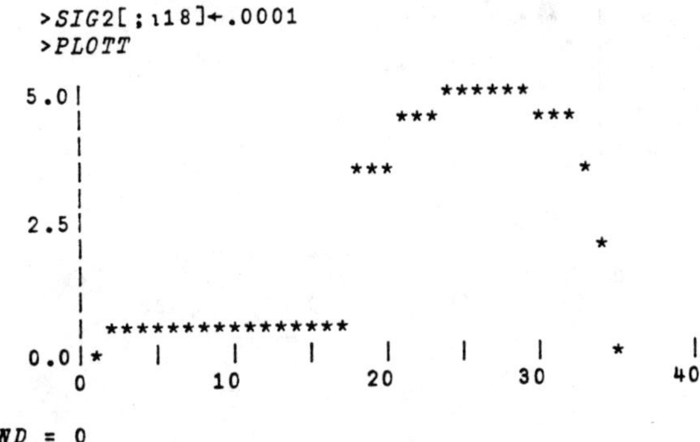

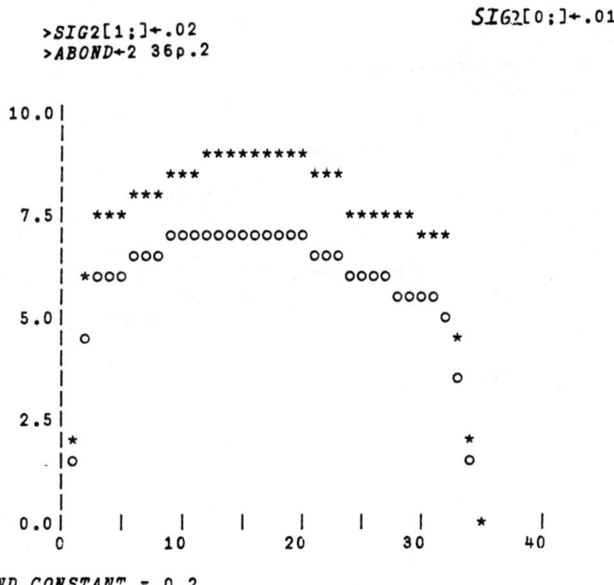

FIGURE 10

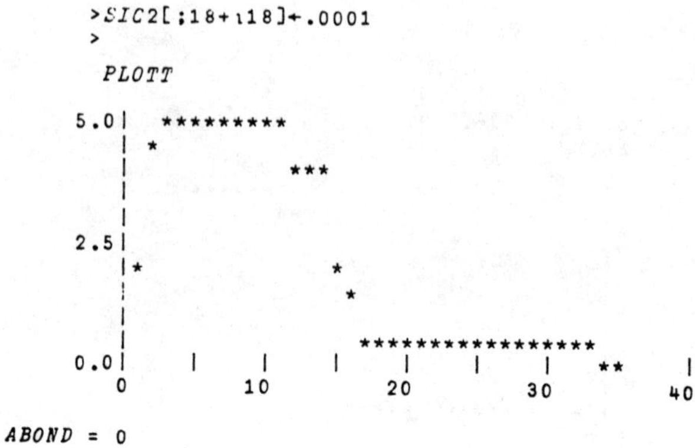

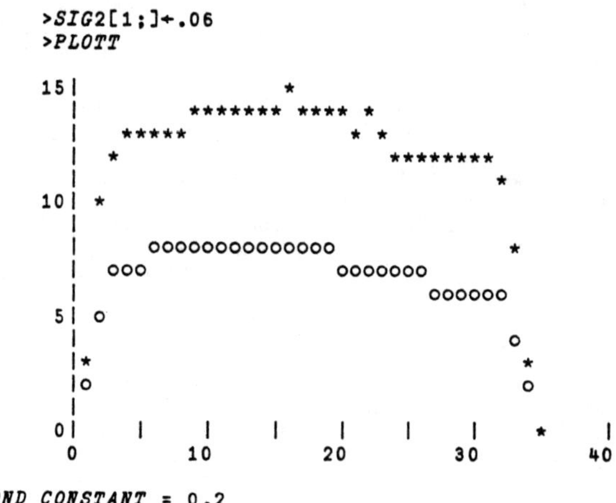

FIGURE 10 (*continued*)

```
>SIG2←2 36ρ.005
>PLOTTT
```

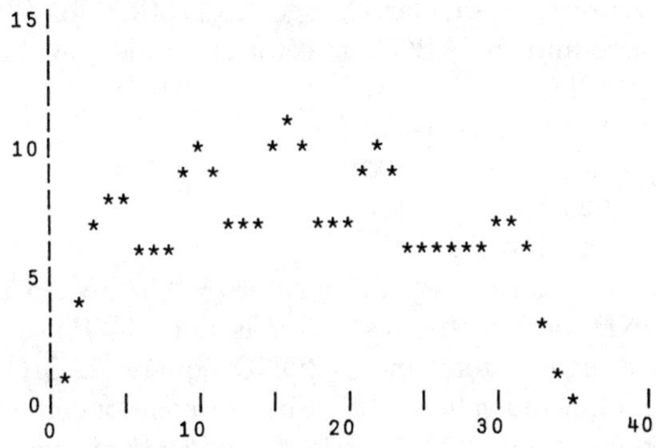


```
ABOND CONSTANT = 0.95
SIG2 CONSTANT = 0.005
```

```
>SIG2[1;]←.001
>PLOTT
```

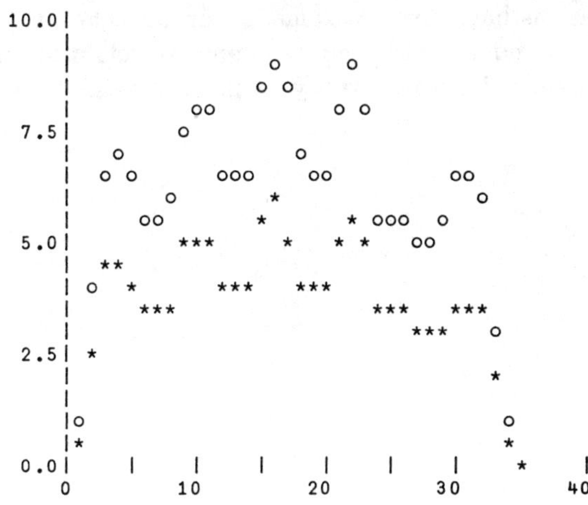


```
ABOND CONSTANT = 0.95
```

FIGURE 11

§4.6. After this excursion we return to the simulations. In each of (a)-(d) of Plate 10 are displayed three independent realizations; SIGMA equals .3 for (a)-(d) and the SIG2 and ABOND vary. Plate 10(c) has the SIG2 vector equal to .01 and the ABOND vector equal to 6; Plate 10(a) has SIG2 as .005 and ABOND as .6; Plate 10(d) has SIG2 as .005, ABOND as .2; Plate 10(b) has SIG2 as .001,, ABOND as .2.

In Plate 11 we illustrate the role of the two rows in the SIG2 matrix. The first one, SIG2[0;], influences scale changes from US(2), while the second one, SIG2[1;], represents angular change from 0(2). In Plate 11(a)-(d) SIGMA equals zero and the ABOND vector is equal to .3 throughout; the SIG2 vector is as follows:

(a) SIG2[0;]←.001,SIG2[1;]←.02
(b) SIG2[0;]←.001,SIG2[1;]←.001
(c) SIG2[0;]←.02,SIG2[1;]←.001
(d) SIG2[0;]←.02,SIG2[1;]←.02.

We also examine another aspect of heterogeneity, when the entries in SIG2 and ABOND vary with sites. This is done in Plate 12. In Plate 12(a)-(d) SIGMA equals zero and ABOND equals .3; SIG2 is a (2,36) matrix with the following values: (a) First 18 components of row 1 equal .02, all others equal to .001; (b) Last 18 components of row 1 equal to .02, all others .001; (c) First 18 components of row 0 equal .02, all others equal .001; (d) Last 18 components of row 0 equal to .02, all others .001. The first 18 components are associated with the right-hand side of the hand, the last 18 components with the left-hand side; row 0 influences scale change while row 1 represents angular change.

The simulations have given us some insight into the functioning of the new global shape model. Although we cannot yet claim to understand its whole complex behavior we are ready to apply it to analysis.

5
Pattern Analysis - Experiments

The synthesis experiments have given us some insight in how the global shape models function and how they are influenced by their parameters. Synthesis experiments should not, of course, be considered as sufficient substitutes for empirical information about the parameters - this should instead be obtained from data as will be discussed in Section 6.

We are ready nevertheless to begin the analysis experiments, in particular structured image restoration. Many sorts of visual noise, degrading the picture, will be considered and the corresponding deformations $\mathcal{D}$'s will differ both in quality and degree.

The purpose is not just to demonstrate that the algorithms work for certain picture ensembles; we want to *find the limits of their applicability.* For this reason we shall make the degradations more and more severe, introducing one type of visual noise after another, *until the algorithms fail and requires modifications.*

The algorithms derived in Section 3 assume one of several $\mathcal{D}$'s. It is natural to start to see how the restoration algorithms work for those deformation studied in 5.1.2.

But this is not enough: to study the robustness of the methods we shall introduce other types of visual noise whose probabilistic structure is essentially unknown and is not simulated but natural.

This will be done in two ways:

(i) non-transparent matter will be distributed over the light table and the hand put *over* it – this creates visual noise outside the hand in addition to the noise that is present anyway all over the picture.

(ii) a transparency with non-transparent matter glued to it is placed *between* the camera and the light table. This creates noise all over the picture.

Successively making the noise more severe we shall *push the algorithms to the point when they start to perform poorly.* The knowledge thus gained will be valuable for further refinements of the methodology presented in this paper.

§5.1. Let us start with a picture that is as clear as we could make it in our laboratory. It is shown in Plate 2(a). Plate 13 displays a histogram of its intensity values and a horizontal profile through four fingers.

The histogram has two well separated peaks as can be expected since the S/N ratio is large here. The lower values corresponds to the inside of the hand, the higher ones to the rest of the picture. In the profile we can clearly see the fingers as troughs in the graph.

The picture, taken at night with all light sources turned off except for the illumination on the screens of the AT and the monitor, has a much better S/N ratio than later on when we introduce massive degradations. Inspection of the values in the matrix below that represents $I^{\mathcal{D}}$ shows, however, more variability than one would guess from looking at the picture itself. Our own visual processing eliminates such effects.

```
24 28 27 29 34 59 89 24 78 83 93 85 81 60 33 27 25 24 26 22
30 30 29 34 31 45 93 85 83 87 89 88 82 48 33 29 26 30 29 29
25 26 27 29 31 40 90 79 84 88 89 85 84 43 34 22 26 29 26 31
27 22 25 25 26 38 84 82 82 84 86 88 84 37 28 30 29 30 25 22
24 30 28 28 29 34 87 86 91 91 86 91 93 38 32 30 29 28 28 23
26 26 21 25 26 36 86 89 90 89 94 90 91 33 29 22 26 28 26 28
26 27 25 21 26 32 70 91 90 81 93 90 75 33 24 25 22 29 21 24
26 27 24 26 29 30 49 95 86 85 89 93 58 33 27 29 20 24 29 28
25 26 22 23 27 29 35 88 89 84 91 93 47 32 26 27 26 22 26 24
25 29 28 27 24 28 27 78 92 88 91 89 35 30 27 30 27 27 26 25
30 24 22 24 27 30 26 38 59 75 81 51 34 31 31 28 25 29 28 29
27 24 27 28 35 28 30 32 38 40 38 34 36 30 25 30 27 26 25 27
25 25 26 22 20 26 24 28 28 33 30 29 28 29 25 30 23 22 24 22
31 30 28 22 28 26 26 32 31 29 31 28 30 26 26 30 28 25 24 23
31 26 29 25 27 27 27 24 31 29 30 28 26 26 29 25 26 28 23 26
29 24 26 28 26 23 22 23 24 29 24 26 24 29 28 28 24 22 22 26
>
```

Applying GLOBAL3 via the driver HANDREST3, we follow the restoration as the algorithm carries out a sweep strategy with 12 sweep areas given by the rows of the matrix

SITES

1	5
3	8
5	11
8	14
11	17
14	20
17	23
20	25
23	32
27	33
29	35
32	0
35	1
0	2

In the matrix each entry means the number of the hinge in the n-gon TEMPLATE1, here $n = 36$. We do not know much about how to design efficient sweep strategies. Of course each hinge must appear at an interior point of some sweep area so that it is given a change to bend. It also appears plausible that in a multi-stage model the earlier stages should have larger sweep areas than the later ones, and this was done in some of our experiments.

In Plate 14 we show the way the boundaries change as iterations proceed, starting with the template in (a) and with the end result in (b). In (a) the red curve is the initial template and the green curve is the new template after stage one; the green curve is the result of GLOBAL2. Plate 14(c) displays this new template (green) the result after 3 sweeps (in blue) and the result after 6 sweeps (in red). The template is reset to this resulting boundary (red) and the next stage is displayed in (d). The red curve being the template and the green is the result after 6 sweeps. In (b) are displayed the initial template (in red), the resulting realization after both stages (in green) and the true boundary (in blue). The restoration is nearly perfect as we expected. We also show, in Plate 15, an example of how the trial arcs of boundaries can look; red is the selected arc.

To make it harder for the algorithm we now add i.i.d. $N(0, \sigma^2)$ noise and show the result in Plates 16-18. Three restorations using GLOBAL3 are displayed for each, Plate 16(a) is a relatively clean picture, CDAN; restorations are in Plate 16(b) with true boundary (blue). Plate 16(c) is a noisy picture with $\sigma = 30$; restorations and true boundary (blue) are in Plate 16(d). Similarly, Plate 17(a) and (b) are for $\sigma = 60$, Plate 17(c) and (d) are for $\sigma = 90$, Plate 18(a) and (b) are for $\sigma = 120$, and Plate 18(c) and (d) are for $\sigma = 150$. the last picture, $\sigma = 150$, seems to have pushed the algorithm to its limit; one of the three realizations is drawn in red to show its erratic behavior.

Plate 19 displays the histogram and a profile for CDA6. Note how flat the histogram is. Plate 20 is the same for CDAL: even flatter histogram, so that it is not surprising that the algorithm failed in the case of CDAL.

We shall use the following conventions when showing pages with four pictures. Then
(a) = noisy picture
(b) = thresholded picture and averages for GLOBAL3 (in green) and GLOBAL4 (red).
(c) = noisy picture and restoration with GLOBAL3 (green or blue)
(d) = noisy picture and restoration with GLOBAL4 (red)

But this was too easy. We now introduce natural visual noise of type (i).

§5.1.1: SAGO. The noisy picture Plate 21 gives the restorations in (c) using GLOBAL3, which is nearly perfect, and with GLOBAL4 in (d), almost as good.

§5.1.2: RUFUS. This noisy picure is shown in Plate 23(a) together with the restoration by GLOBAL3 in (c); not too good. GLOBAL4 performs better, see (d) and the averages in (b). Noise of this type is well handled by local estimation. In Plate 22 we display the thresholded versions with different choices of threshold constants. It is impossible to find a good constant, in the sense of both suppressing noise and keeping most of the shape intact.

§5.1.3: COINS. GLOBAL4 was fooled once by this noise, see Plate 24(d); GLOBAL3 worked well as shown in (c). It is not quite clear why the incorrect decision in one of the three boundaries in (d) was made. The averages of the two GLOBAL's are in (b).

§5.1.4: WRICE. This picture is better handled by GLOBAL4, see Plate 25(d), than GLOBAL3, see (c), which over estimates three of the finger tips and also a bit of the thumb. It shows the strength of local estimation well. Averages in (b).

§5.1.5: PAPER1. The long correlation noise in Plate 26 did not confuse GLOBAL3, see (c), although some uncertainty appears in the decision at the tip of the ring and long fingers. This is surprising, we had expected failure with global estimation. GLOBAL4 is shown in (d) and the averages from GLOBAL3 and GLOBAL4 are shown in (b).

§5.1.6. PAPER2. Now we try to seriously confuse the algorithms by placing an object, somewhat finger like, between the thumb and index finger in Plate 27(a). This actually happened, but not consistently for

GLOBAL3. Indeed, two realizations shown in (c) were wrong but two in (b) were correct, actually very good.

On the other hand GLOBAL4 performed poorly, see (d); one realization missed the little finger, and one self inserted which seldom has happened otherwise. This is for the same reason as in 5.1.5. Here we have reached a limit of applicability of the algorithms in their present form.

§5.2. We now go ahead to visual noise of type (ii), which often presents a greater challenge to the algorithm. We start with the easier ones of them.

§**5.2.1: PEAS.** Here GLOBAL4 did an excellent job considering the noise behavior, see Plate 28(d), although the thresholding cannot suppress this sort of noise see (b). The noise has long correlations and it was not expected that the algorithm could deal with it but it did. Note that the correlation length is of the same order of magnitude as the widths of the finger; this should cause trouble, but this did not materialize.

§**5.2.2: CLOTH.** GLOBAL3 did a fine job here, see Plate 29(c), and the same is true for GLOBAL4 in (d), although the tip of the index finger is missed for some reason that we do not understand. In (b) we show averages for each algorithm together with the thresholded picture. Note that the main direction of the noise is about the same as that of the fingers. This would tend to make it difficult to find the orientation (but not the length) of the finger but the algorithm coped well.

§**5.2.3: PLBAG.** About the same is true for this picture in Plate 30. The dark region across the hand makes GLOBAL3 slightly overestimate the width of palm, see (c). In (b) we show the results of both GLOBAL3 and GLOBAL4. Perhaps local estimation performed a little better than the other one.

§**5.2.4: SCARF.** GLOBAL3, in Plate 31(c), performs well except for some uncertainty at the tip of the ring finger, GLOBAL4, in (d), is even more successful. In (b) the respective averages are shown.

This noise is handled without problems.

§**5.2.5: GLUE.** This picture has some interesting deformation in it. The obviously present blurring should perhaps have been treated with an algorithm that we had developed but this was not done.

Instead we applied GLOBAL3 and the result shown in Plate 32(c) looks great. This was another unexpected success! One of the restorations by GLOBAL4 fails, see (d). Thresholding in (b).

§5.2.6: SWEATER. Plate 33(a) is the noisy picture, (c) and (b) are two thresholdings, and (d) is the histogram of pixel values with just a single peak. GLOBAL3 did well in Plate 34(c) but GLOBAL4 had trouble at the thumb, see (d). This may be because of the noise which is especially severe around the thumb. The averages are in (b) together with the thresholded figure.

§5.2.7: TDAN8. Plate 35(a) has only little noise in it and is well restored by both GLOBAL3 in (c) and GLOBAL4 in (d) . Some variability in the lengths of fingers is as it should be: the posterior probability measure has some, but not very much spread due to noise. Thresholding does quite well, see (b). Plate 36 displays the histogram of pixel values; the histogram has two peaks, perhaps, this facilitates natural restoration. Also in Plate 36 is a profile of pixel values along the line through four fingers; the fingers can be distinguished in the profile.

§5.2.8: TDAN11. We now come to three pictures of the same hand but with decreasing S/N ratios. Plate 37 has a bit more noise than the previous case. In (c) we see the noisy picture with GLOBAL3 restorations, in (d) GLOBAL4 restorations; the true boundary is shown in blue. The restorations are good but not outstanding. Plate 38 displays the histogram of pixel values and also a profile of pixel values along the drawn line. The known boundary is superimposed so that the resulting "spikes" occur when the line crosses the fingers, and are just artifacts. They serve to identify the four fingers.

§5.2.9: TDAN12. Now a noisier picture, Plate 39. Result and true boundary in (c) and (d); true boundary together with thresholded figure in (b). In (c) we see real mistakes, while (d) is fairly good. This is the picture discussed just prior to section 1.1.

Histograms of picture and profile of line given in Plate 40; similar to previous picture.

§5.2.10: TDAN13. Now a really noisy one! GLOBAL3 results and true (blue) are shown in Plate 41(c) and GLOBAL4 results and true (blue) in (d). Only fair or poor performance, the tips of fingers are missed a number of times, and the indentations are wrong. The reason why this happens becomes clear when we look at the thresholded figure in (b): the transitions finger-background is not sharp enough to be. Finally the histogram of the pixel values appears in Plate 42, only one peak is visible so that restoration is hard. Without the "spikes" in the line profile it would be difficult to know where fingers begin and end.

This picture was not handled very successfully, but GLOBAL4 clearly did a better job than GLOBAL3.

§5.2.11. XRAY3. Now let us make it harder for the algorithm. The next picture, Plate 43, has strong variations in background with long correlations, see (a). In (c) we show GLOBAL3, very good except at the thumb, and in (d) GLOBAL4, not quite as good, missing tip of thumb and little finger.

§5.2.12: XRAY2. Plate 44 is similar to previous one, but higher noise contrasts. In (c) and (d) we show the results of GLOBAL3, overestimating fingers, and GLOBAL4, which is better except at the thumb.

Considering the nature of the noise the results can perhaps be judged fair considering the degradation level of the deformed picture. In (b) we display the best thresholding for each of the two algorithms.

§5.2.13: XRAY1. Plate 45 has even more noise. GLOBAL3 is fooled at the little finger and thumb, see (c). GLOBAL4, in (d) performs even worse at the little finger. The algorithms are outside their domain of applicability. In (b) the averages of the two algorithms are superimposed on the best thresholding.

To deal with such noise knowledge of its strong dependence must be built into the algorithms. We have described how to do this Section 3.1, but it has not yet been tried.

§5.2.14: SAGOB. GLOBAL3 and GLOBAL4 restorations are given in Plate 46(c) and (d), respectively. GLOBAL3 is a little confused at the fingertips and GLOBAL4 makes one serious misjudgment of where the indentation is between the index and ring fingers. The algorithm for blurring has not yet been implemented. The averages of the algorithms are shown on the thresholding in (b).

§5.2.15: SAGOD. Much like the previous; with even longer correlation. Thresholding given in Plate 47(b). GLOBAL3 in (c) fails miserably; GLOBAL4 in (d) did an excellent job however.

§5.3. We now illustrate what happens when the illumination has a strong gradient and show this with a single picture Plate 48.

§5.3.1: CHOW. In (c) we show the true boundary in blue and the reconstruction by GLOBAL3, using global estimation. The result is extremely poor, especially at thumb and little finger region. This should be compared with (d), employing GLOBAL4, which works much better, although there is still an area around the little finger that is missed. This proves the need for local estimation in cases with strong non-uniformities in the lighting. In (b) we show the true boundary together with the thresholded figure. Plate 5 displays the histogram of the picture and a profile of pixel values along a vertical line; the profile shows the gradient with a 90% increase from one side to the other.

§5.4. We now examine what happens if a part of the picture is masked out: *image extrapolation.* For the next four pictures only inside the red rectangle is observed. The pictures involve very little noise or synthetic noise and so only GLOBAL3 is applied. The mask used will be shown in quadrant (b).

§5.4.1: MASK1. Part of the relatively clean picture from Plate 3(a) is masked in Plate 49. In (c) are three restorations (green) from GLOBAL3 and true boundary (blue). The average of the three is shown in (d) with the true (blue). The variation occurs where one would expect.

§5.4.2: MASK2. Plate 16(c) with added synthetic noise $\sigma = 30$ is shown in Plate 50. GLOBAL3 results given in (c). A lot of uncertainty at the thumb. The average give in (d) with the true boundary (blue).

§5.4.3: MASK3. Plate 17(a) with added synthetic noise $\sigma = 60$ is given in Plate 51. GLOBAL3 restorations (green) in (c) with true boundary (blue). The average of the three restorations is shown (green) along with true (blue) in (d).

§5.4.4: MASK4. A larger mask is applied to the picture with synthetic noise $\sigma = 90$. Three restorations (green) are shown in Plate 52(c), (d), and (b) with true boundary (blue). The algorithm failed miserably. The reason being the rectangle of observability was too small; almost all the observed is inside the hand.

§5.5. Reduced Resolution. The next two figures are reductions of the 128×120 picture TDAN11 discussed in 5.2.8. GLOBAL4 algorithm was applied.

§5.5.1: TDAN11R1. Plate 53 displays the reduction of (a) to a 64×60 in (b), 32×30 in (c) and 16×15 in (d). Plate 54 gives three restorations from GLOBAL4 (red) for each resolution along with true boundary (blue). The algorithm did quite well; even at the resolution 16×15 it seems to do what human judgement might suggest.

§5.6. Pathologies. In the following three cases we create pictures relatively noise free but with the hand modified.

§**5.6.1: LEFT.** In Plate 55(a) a paper attachment was made to thumb. GLOBAL3 and GLOBAL4 are shown in (b) and (c) respectively. Averages in (d). GLOBAL4 didn't do well, but that is not surprising for this type of noise. GLOBAL3 did as well as one could expect. The pathology detector, discussed in Section 3.2, would detect that the thumb was out of proportion and this was the main objective here.

§**5.6.2: RIGHT.** Same as previous but paper attached to right side of hand. GLOBAL4 in Plate 56(c), GLOBAL3 in (b), averages in (d). Moderately successful.

§**5.6.3: MISSING.** Plate 57(a) has index fingers shortened. *A pathology detector*, using the fact that the s-matrices on the edges about the index finger are being forced to scale down, detects the pathology.

Color Plates

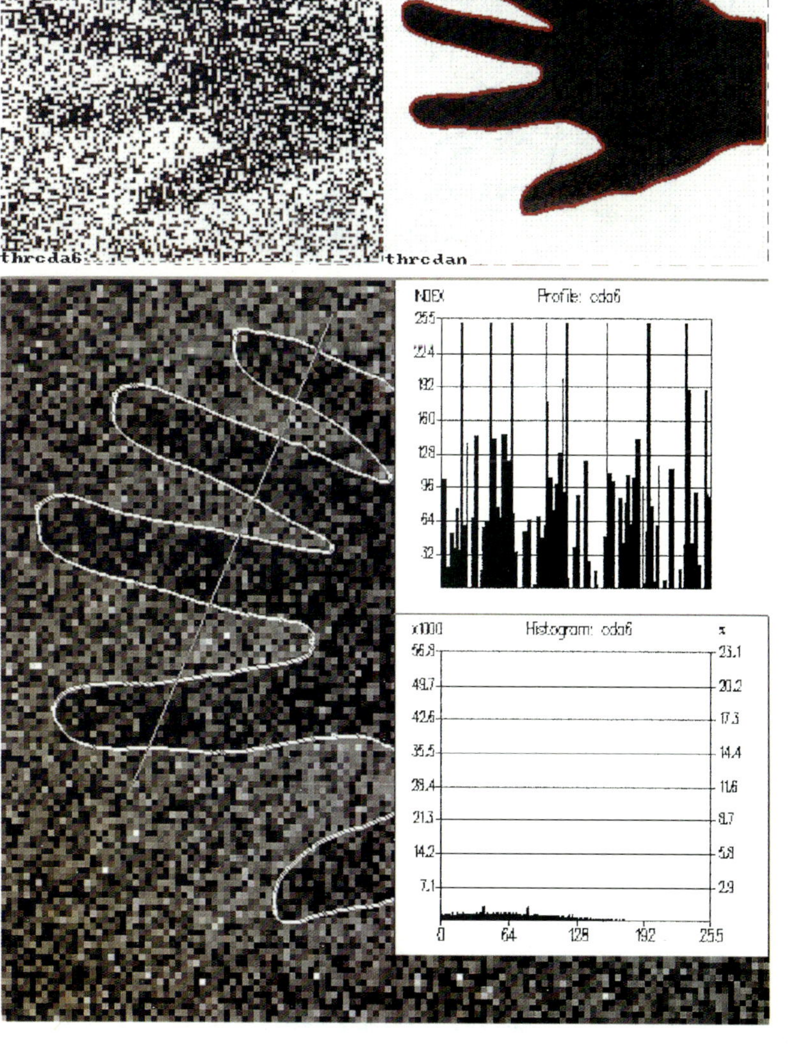

Plate 1

Plate 2

Plate 3

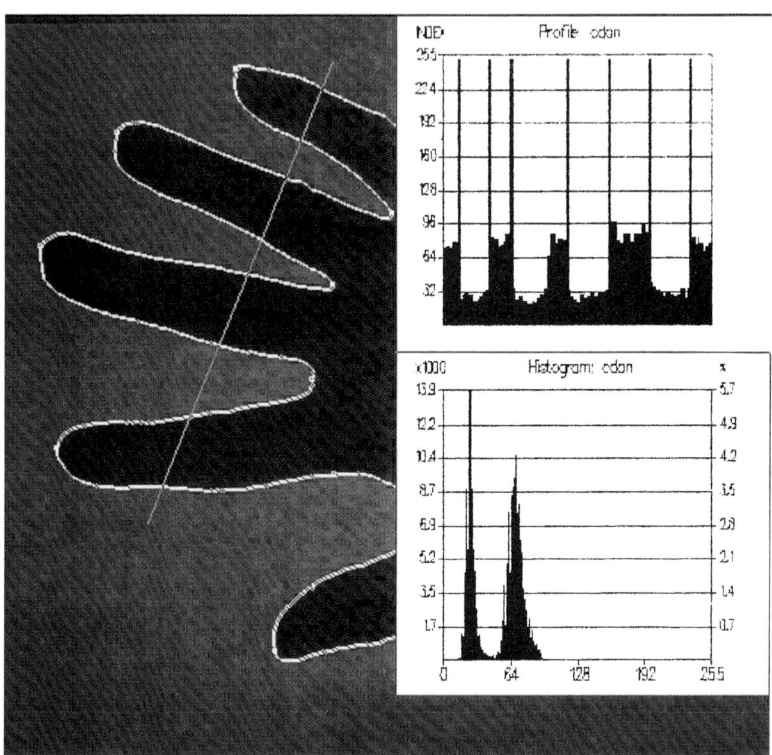

Plate 4

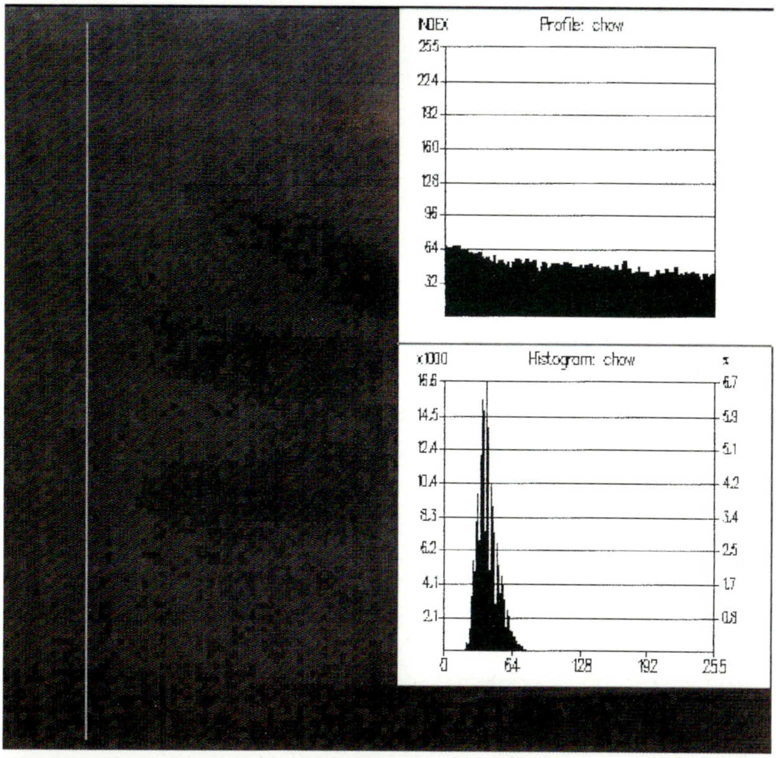

Plate 5

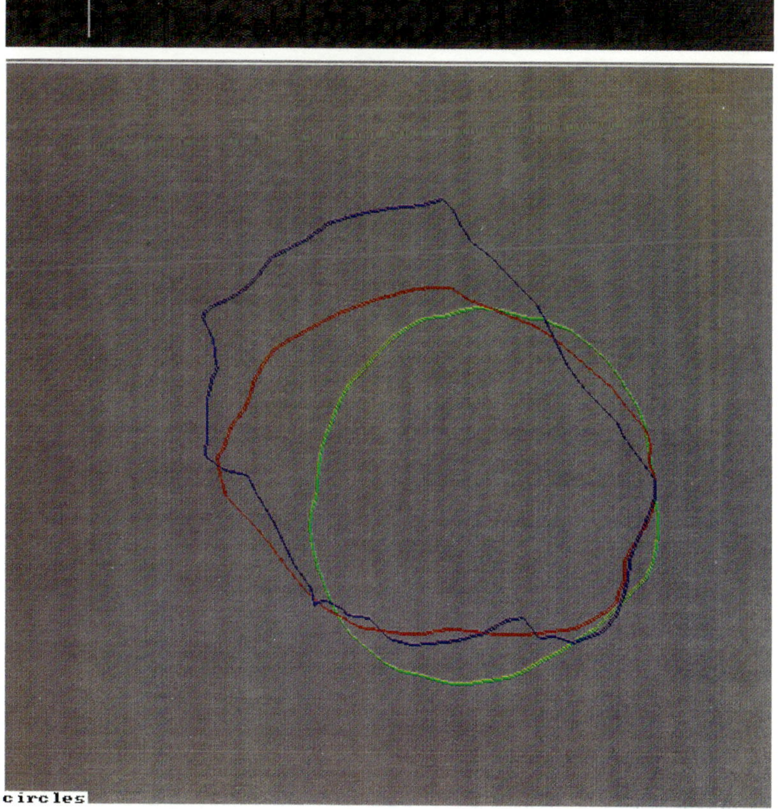

Plate 6

Plate 7

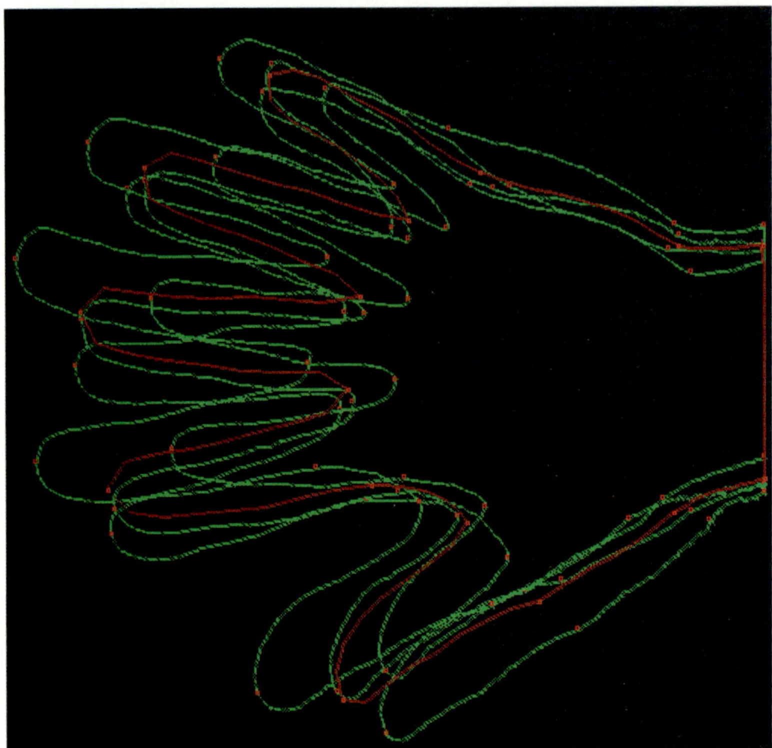

Plate 8

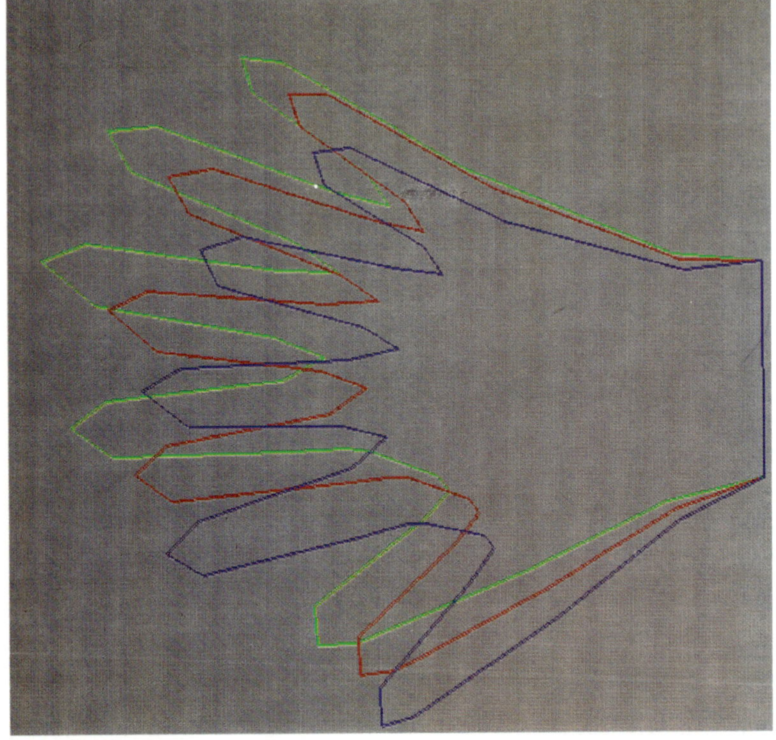

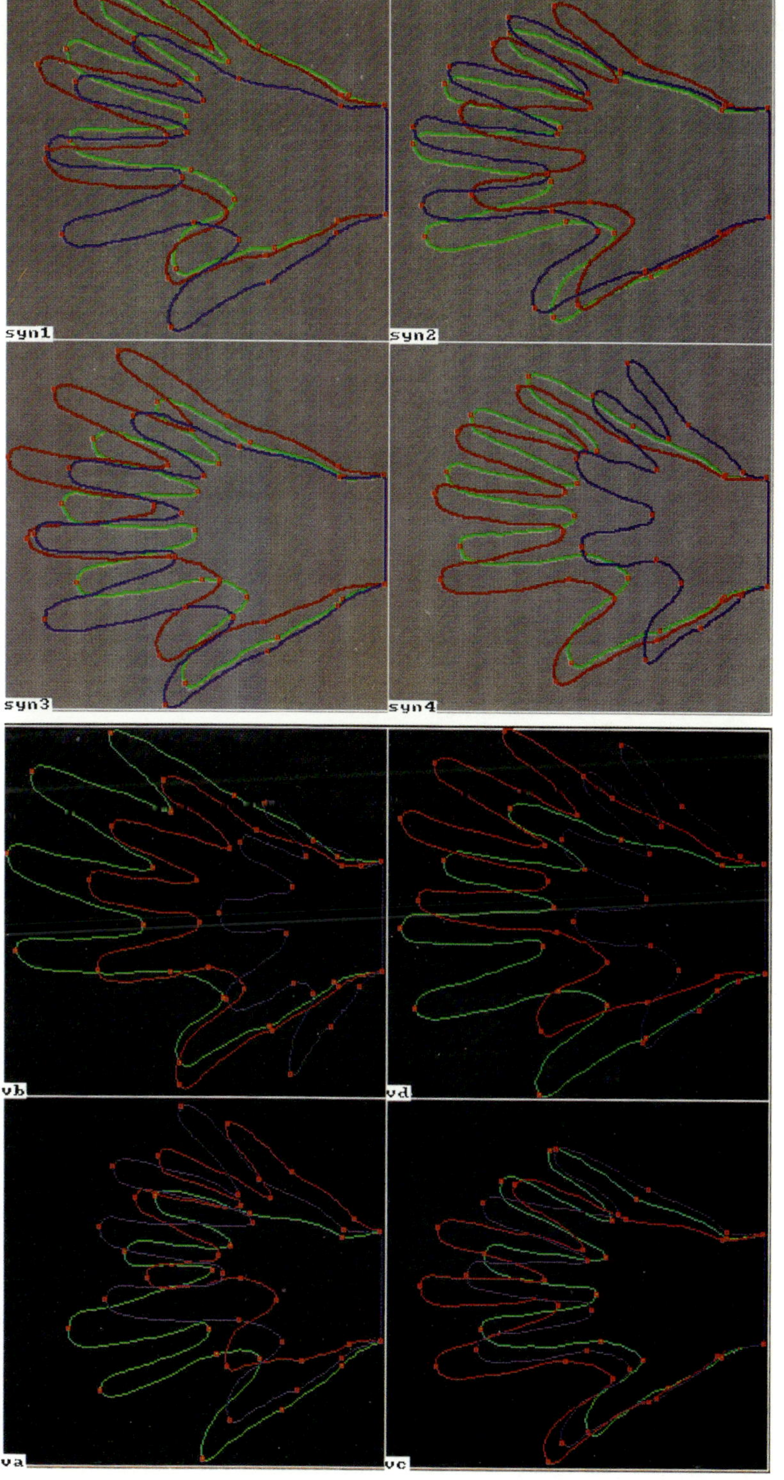

Plate 9

Plate 10

Plate 11

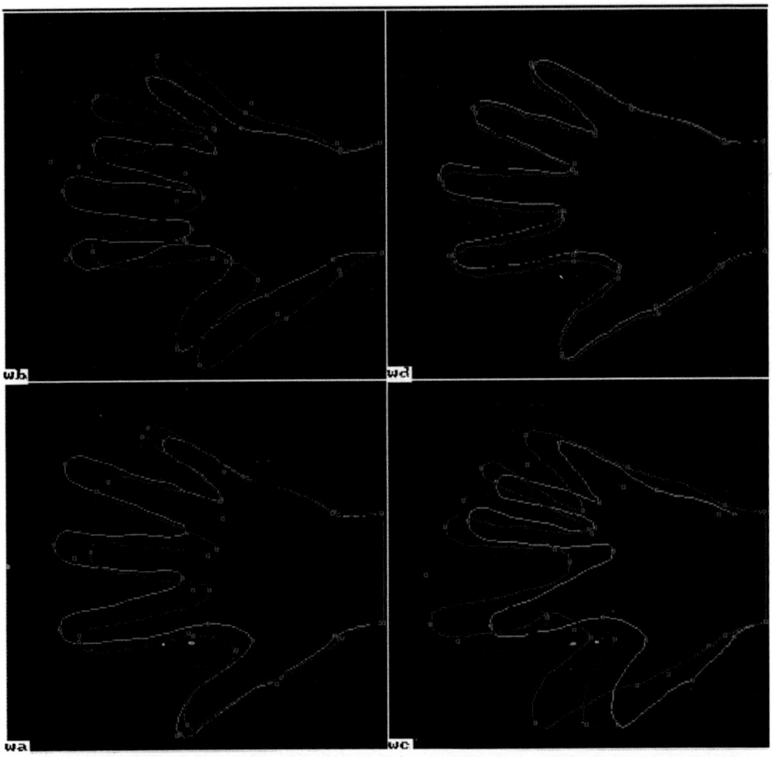

wb

wd

wa

wc

Plate 12

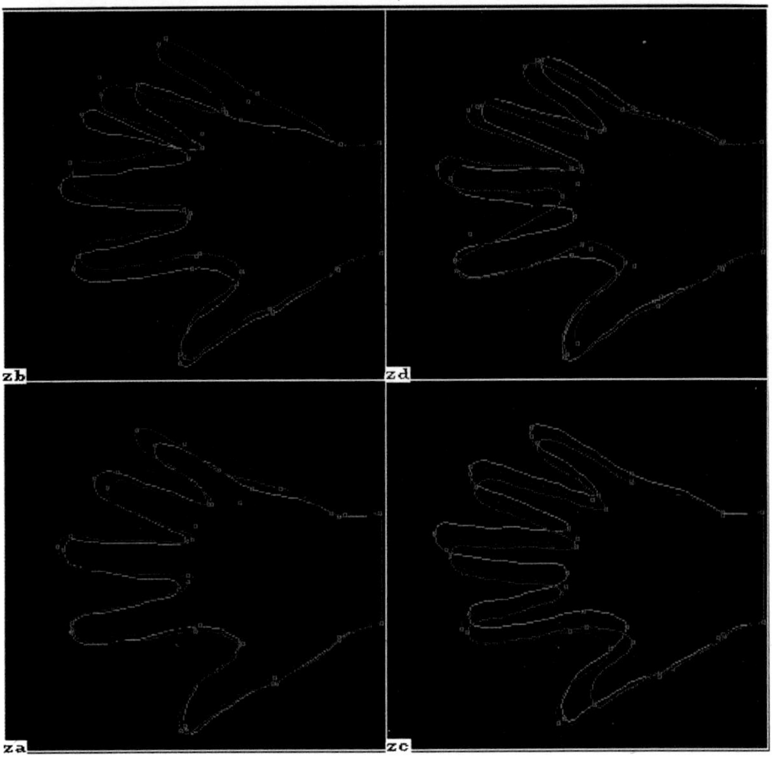

zb

zd

za

zc

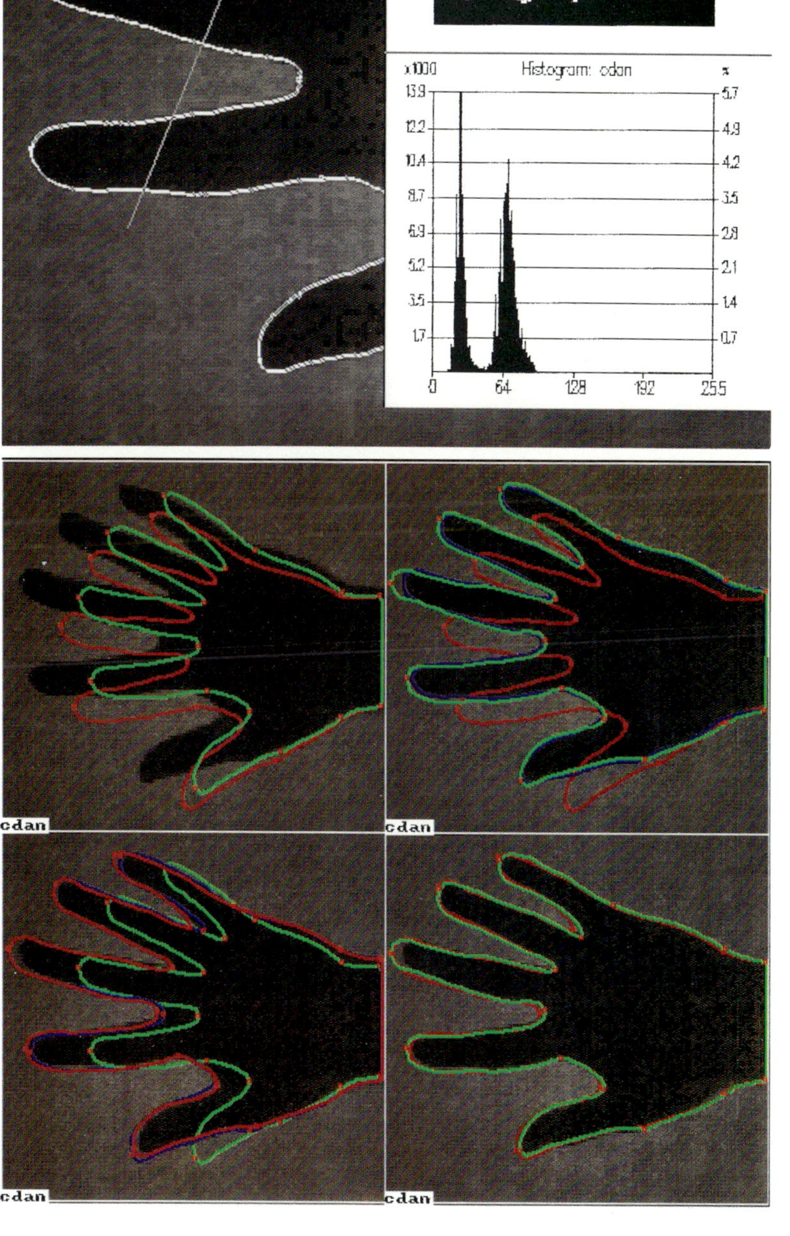

Plate 13

Plate 14

Plate 15

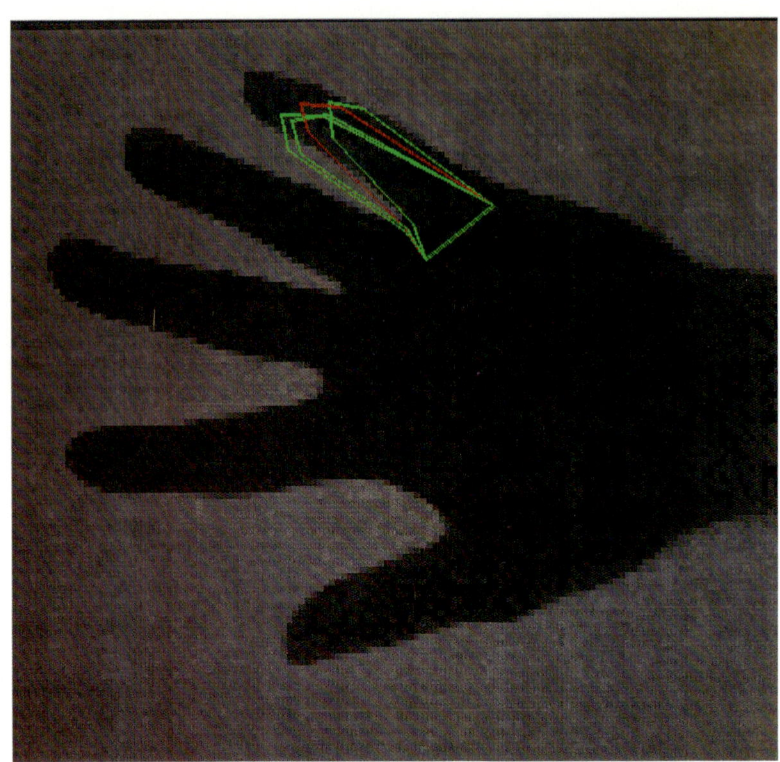

Plate 16

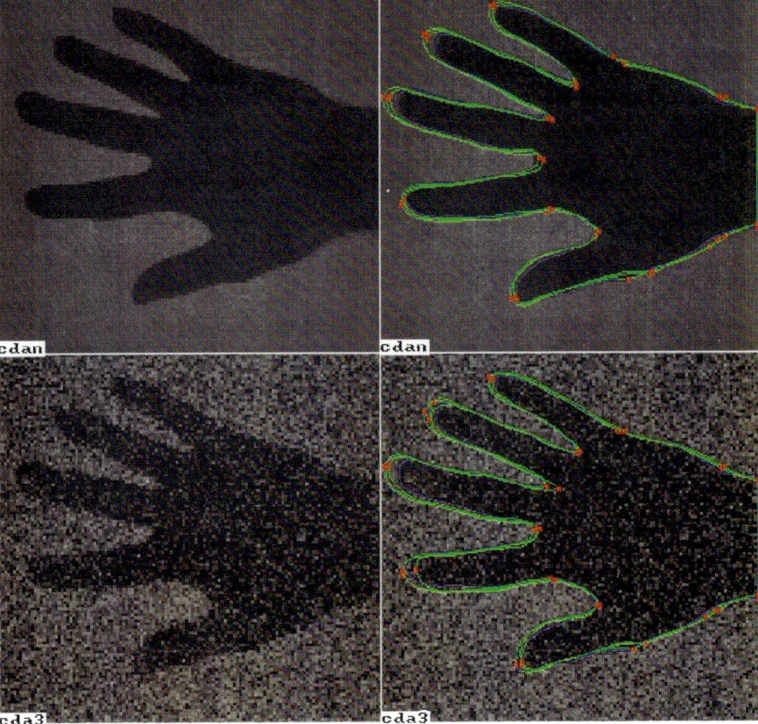

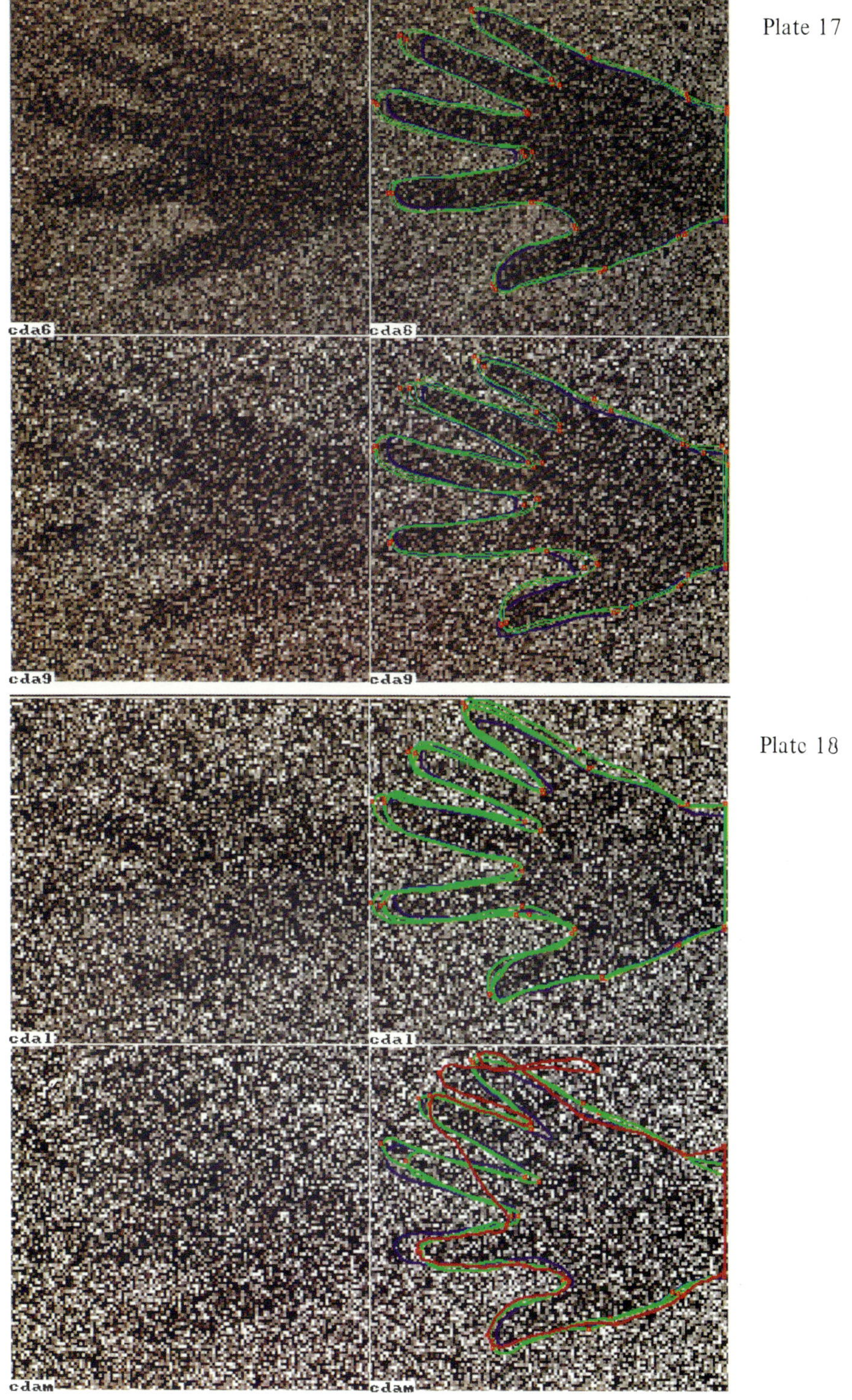

Plate 17

Plate 18

Plate 19

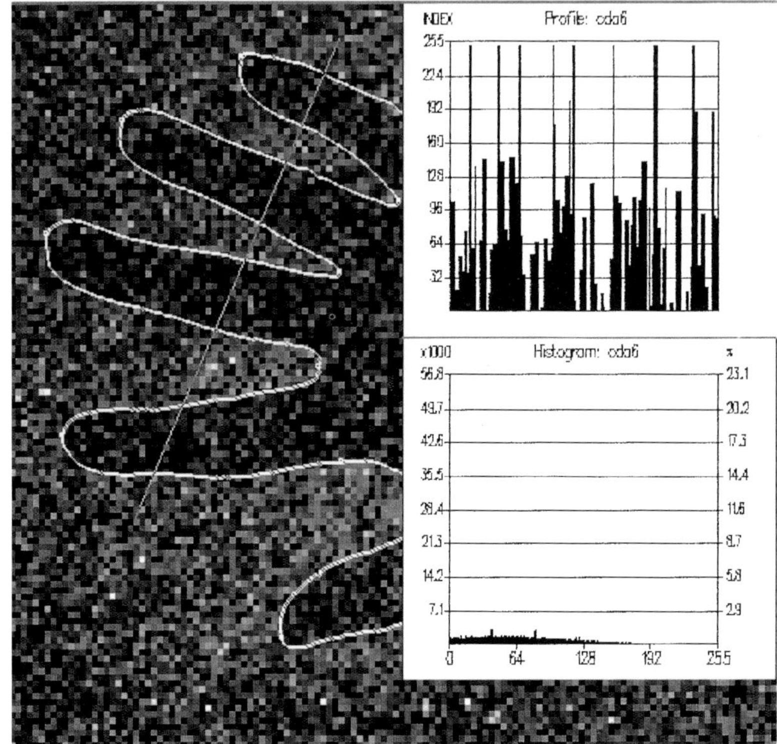

Plate 20

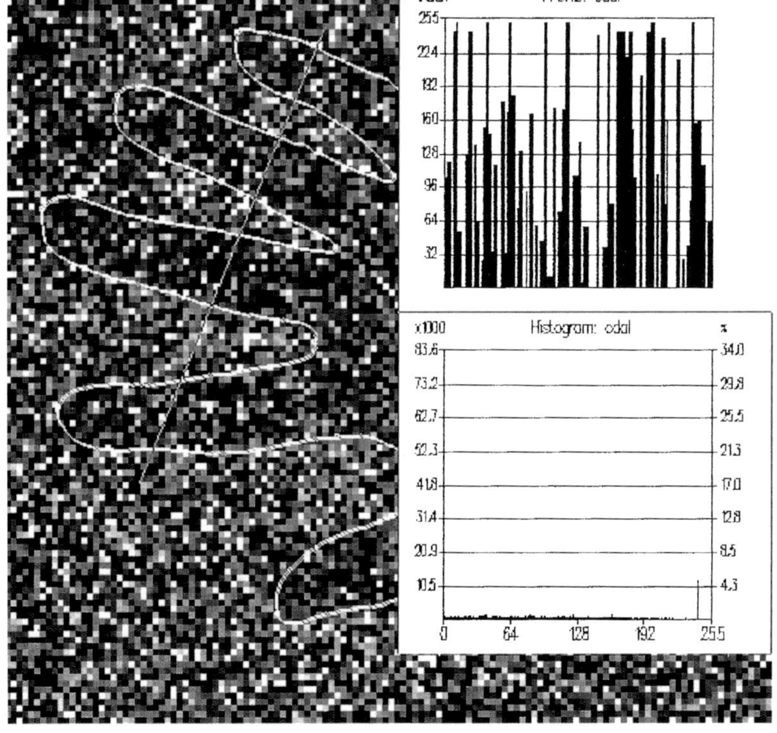

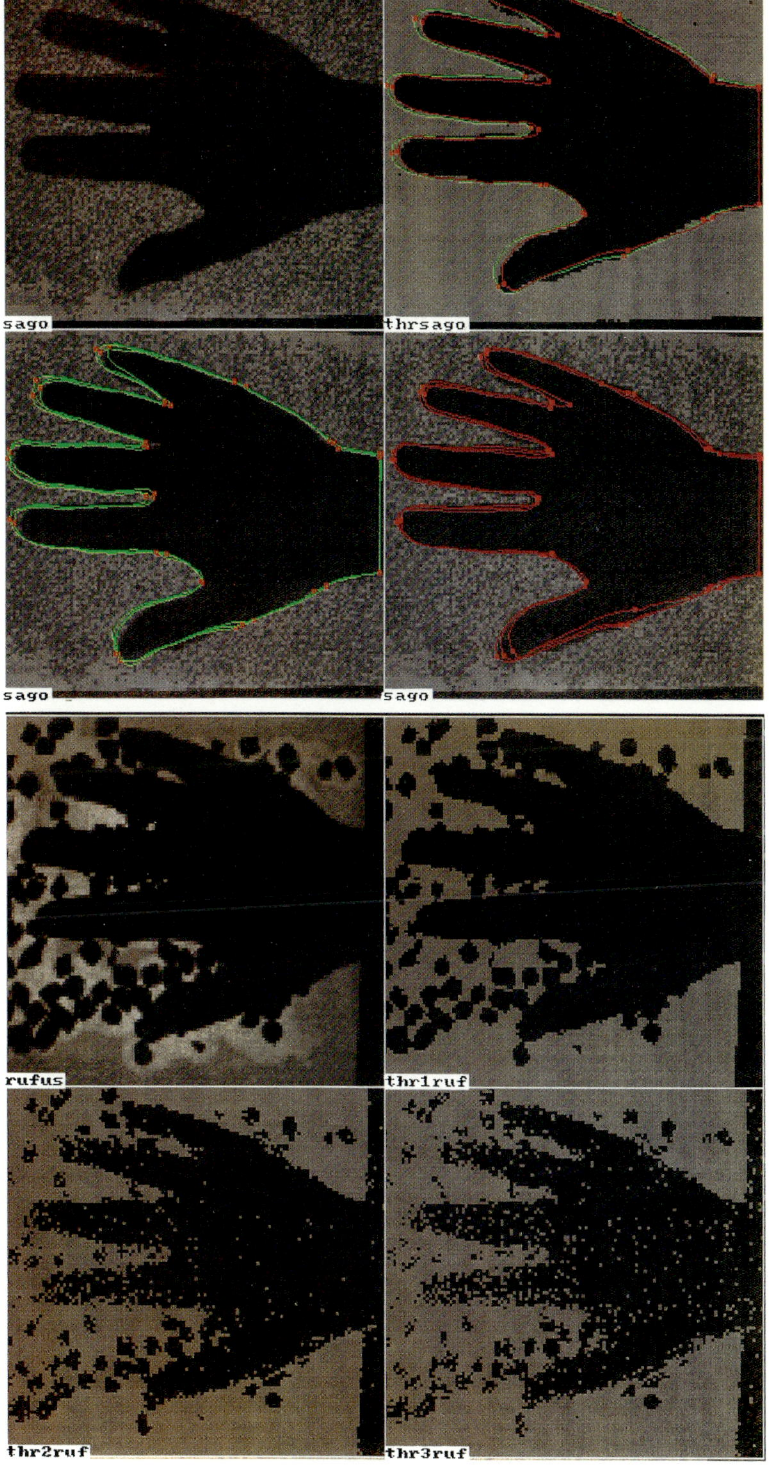

Plate 21

Plate 22

Plate 23

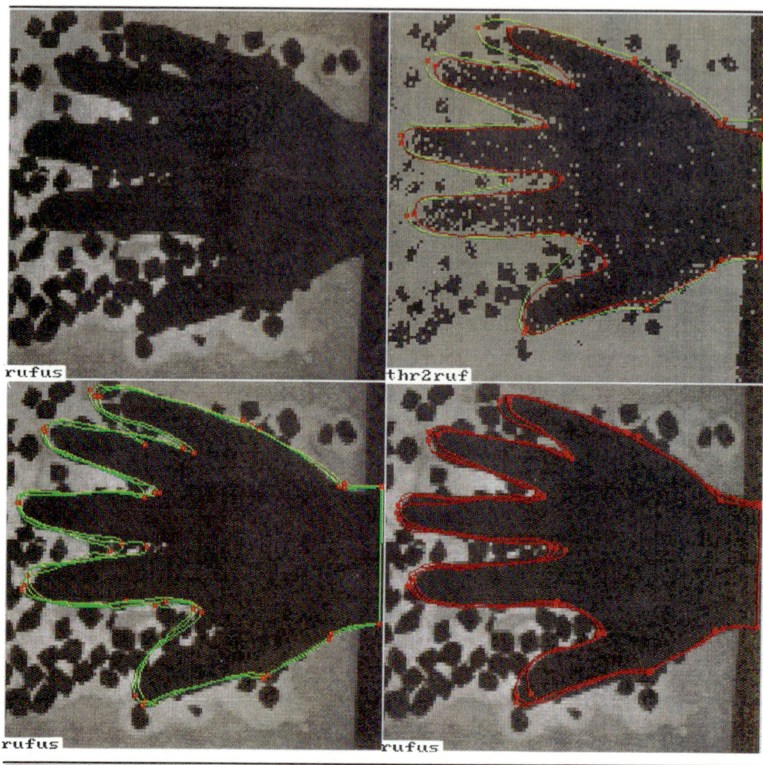

Plate 24

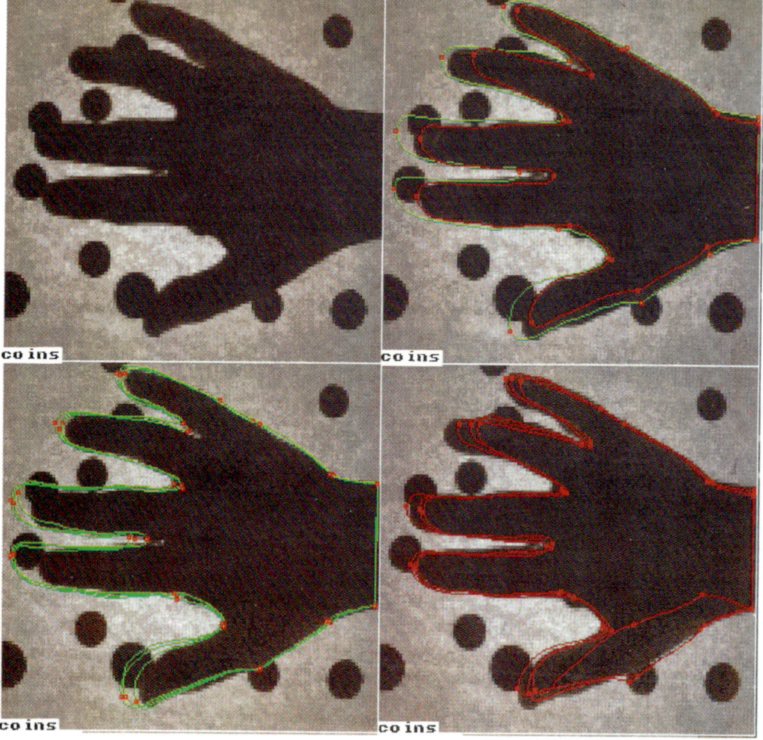

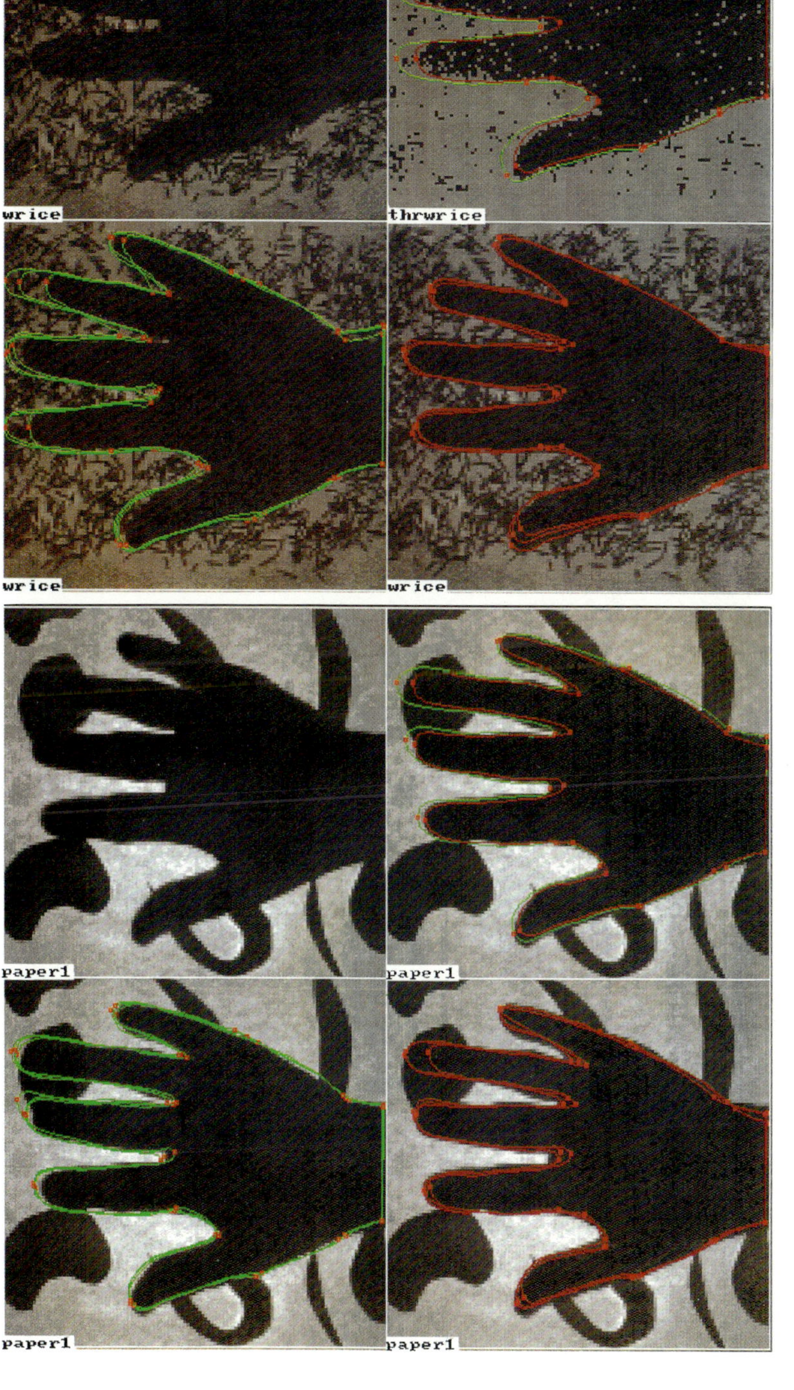

Plate 25

Plate 26

Plate 27

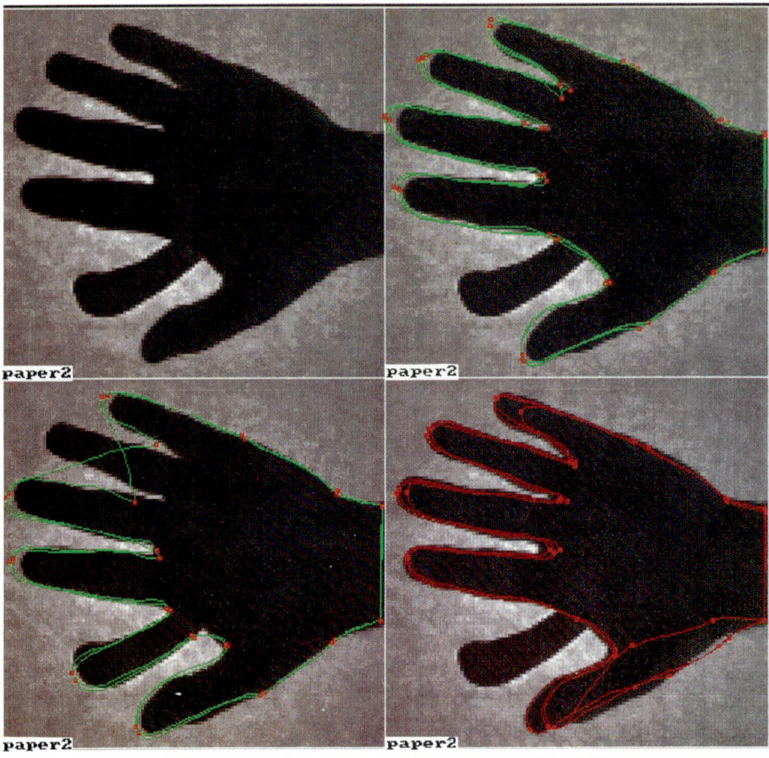

Plate 28

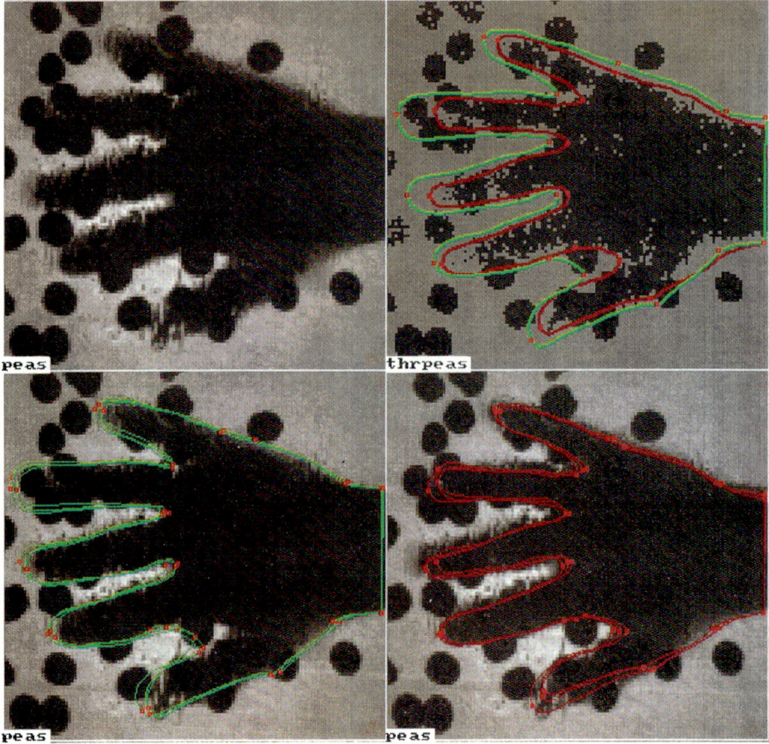

Plate 29

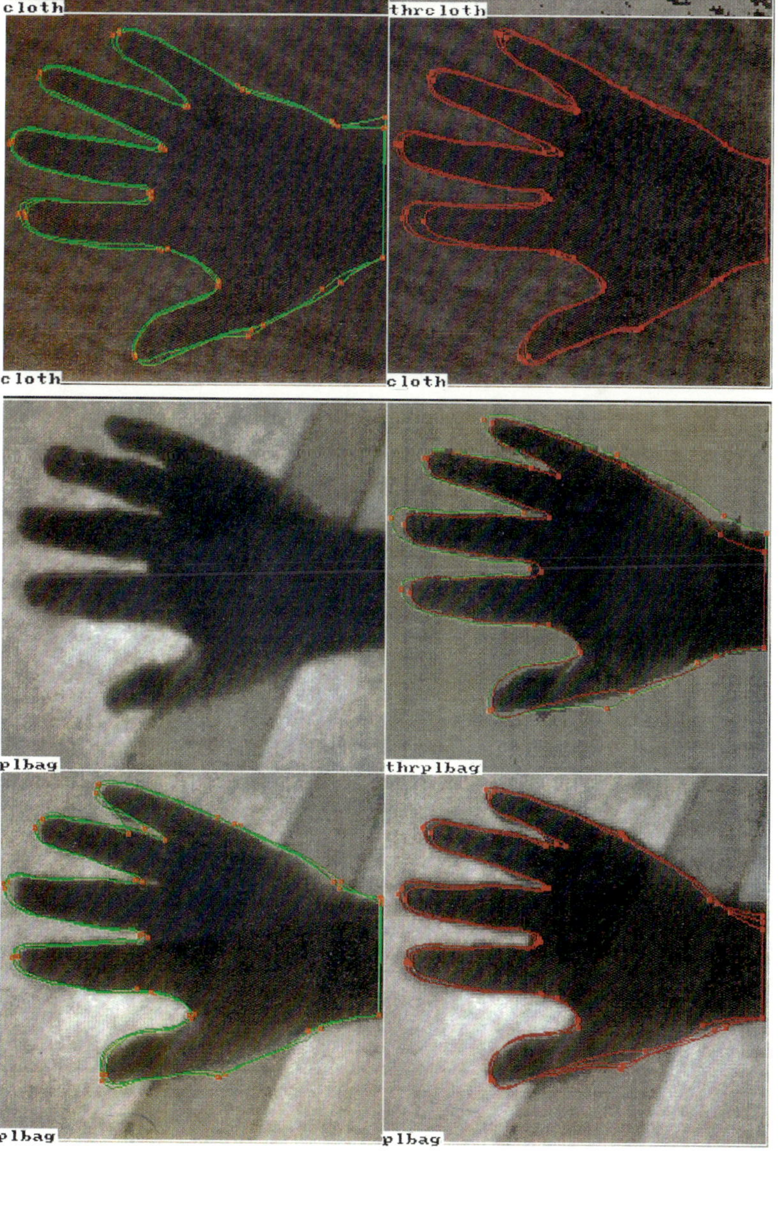

Plate 30

Plate 31

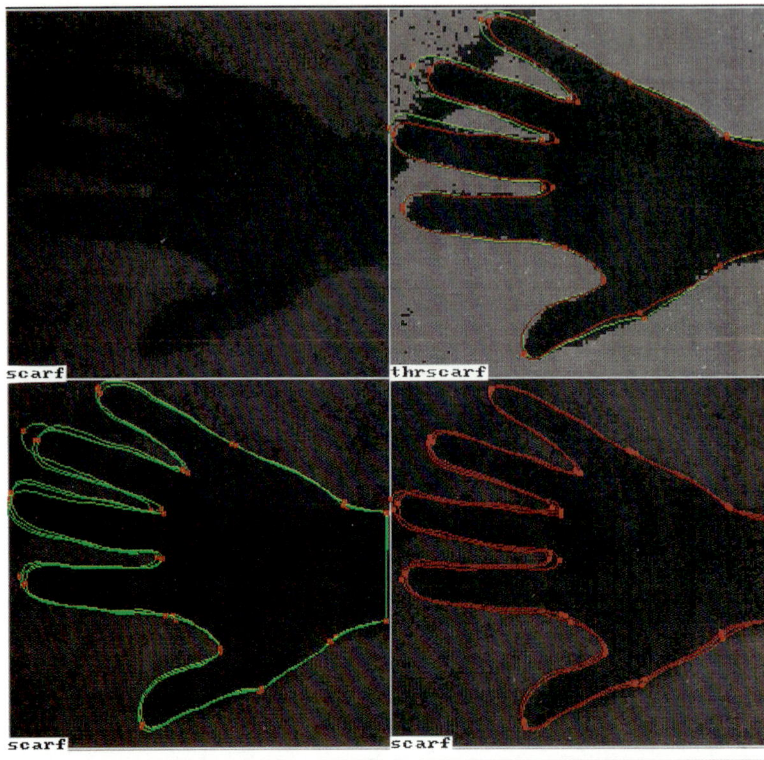

Plate 32

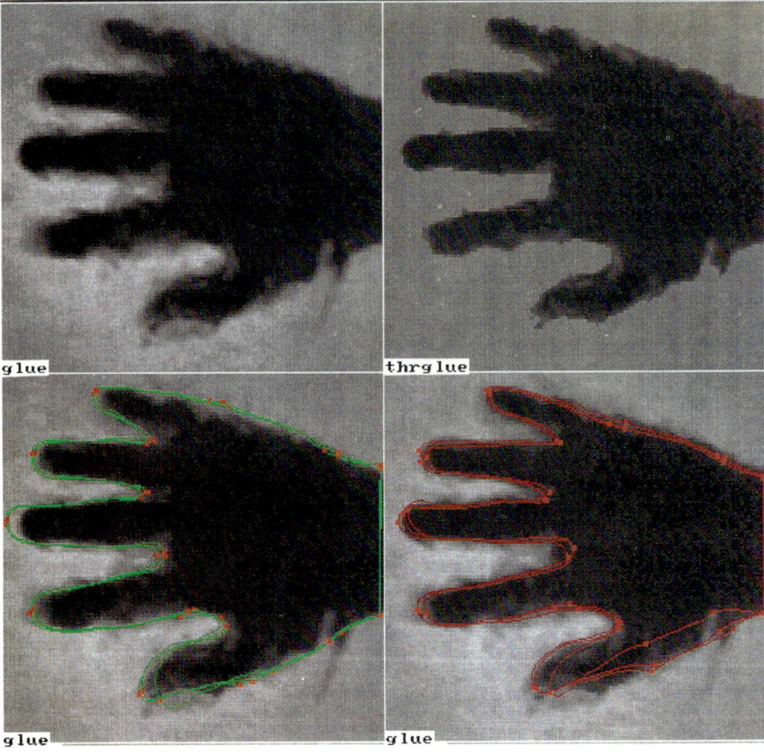

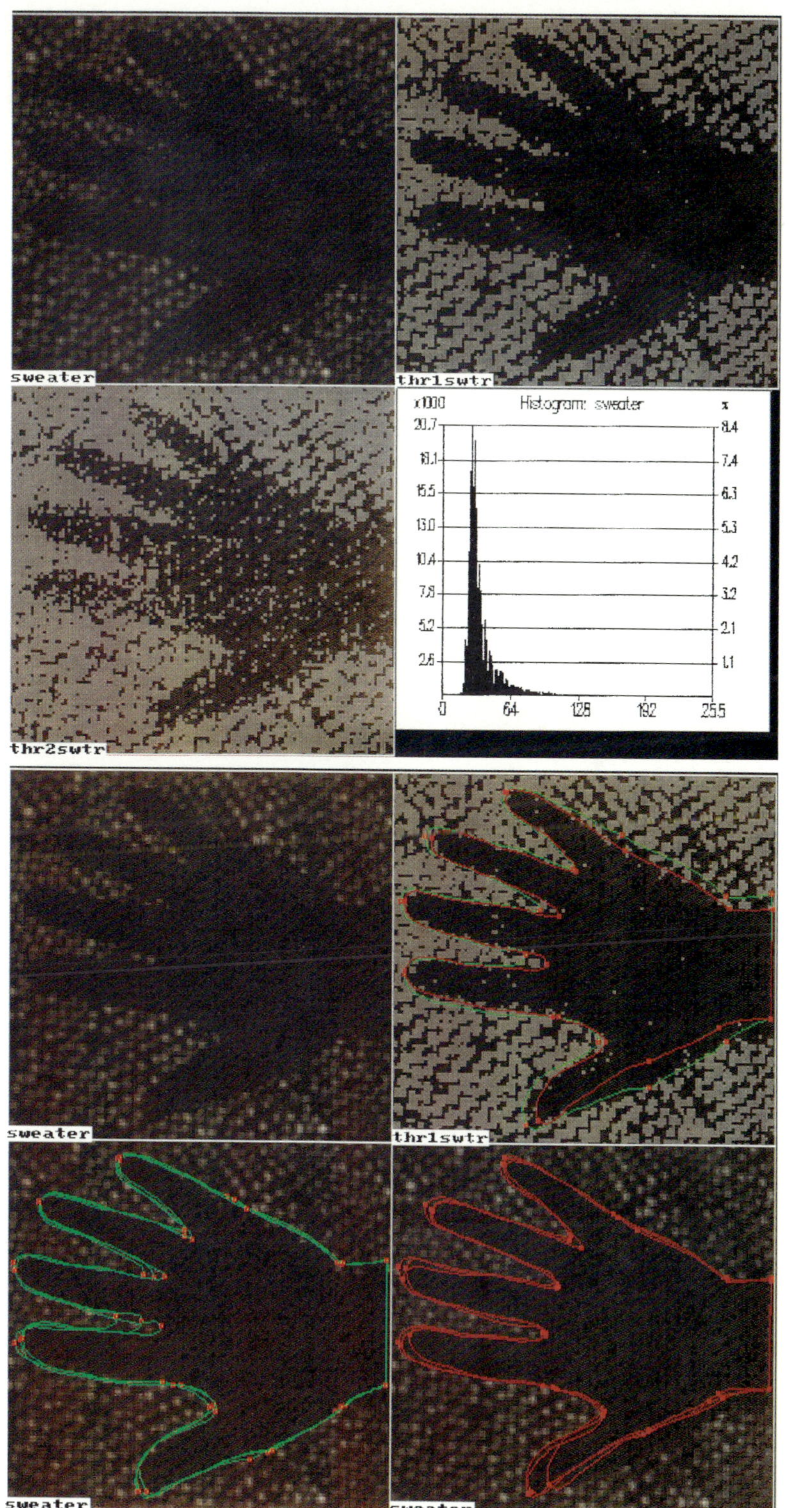

Plate 33

Plate 34

Plate 35

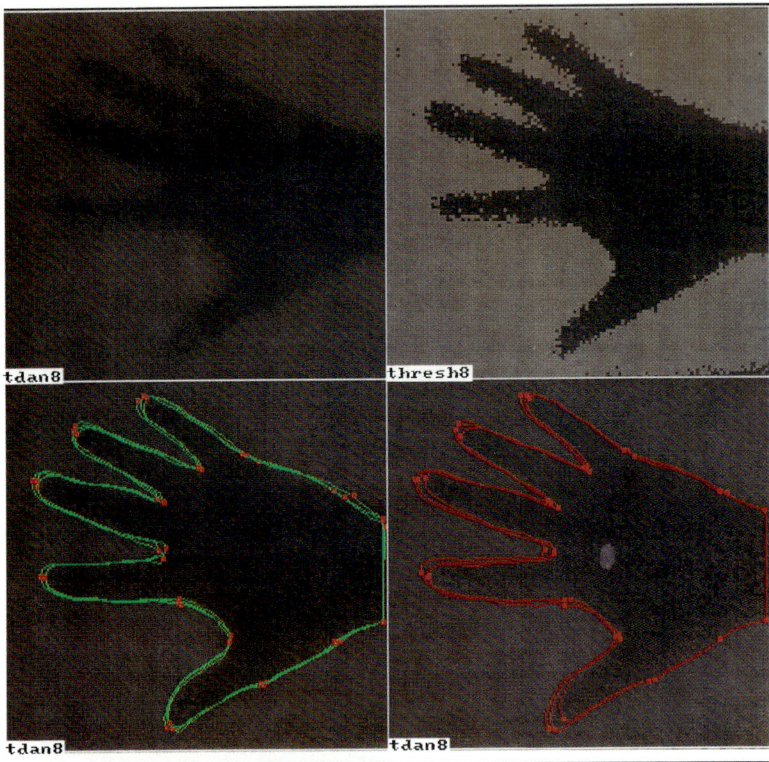

Plate 36

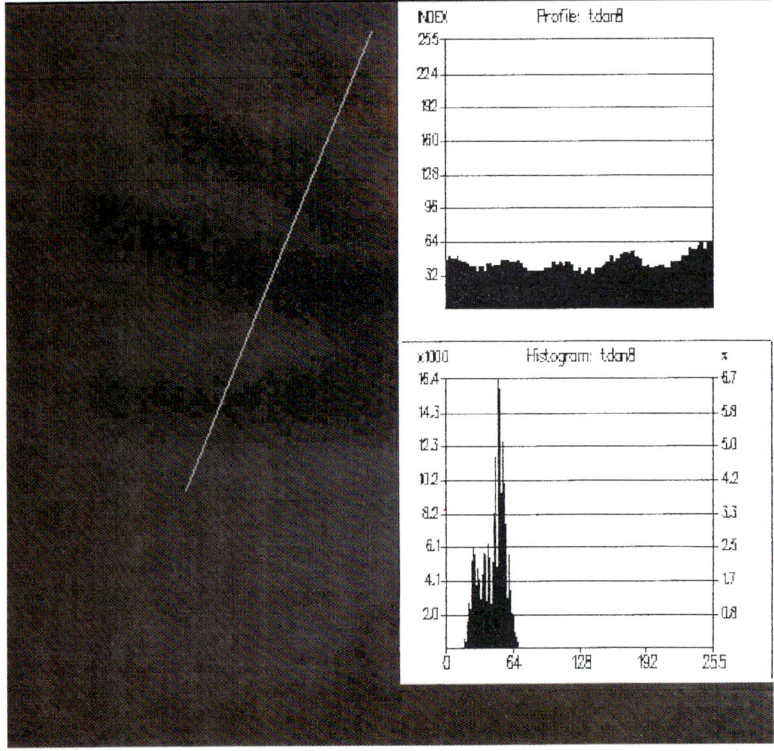

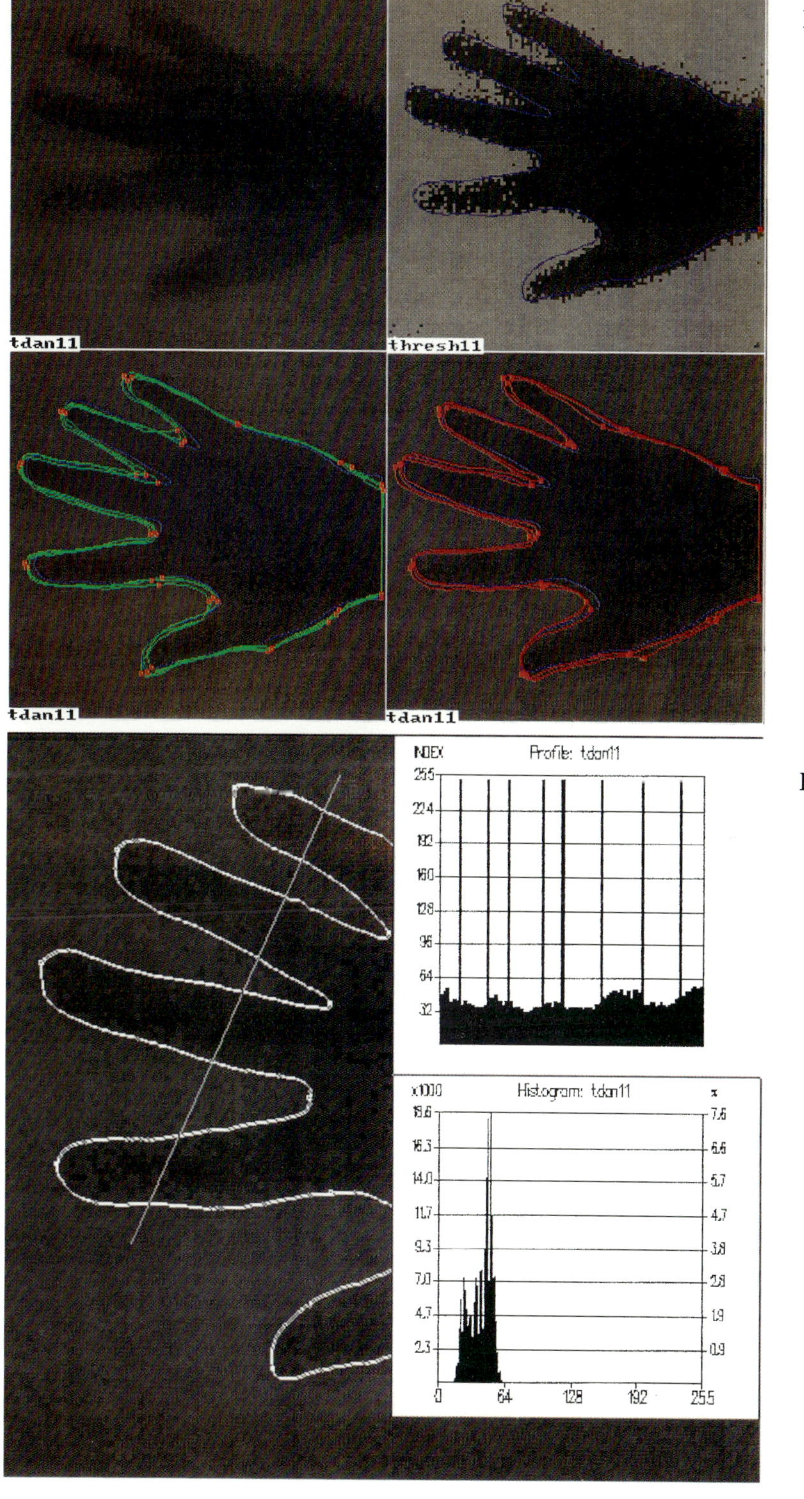

Plate 37

Plate 38

Plate 39

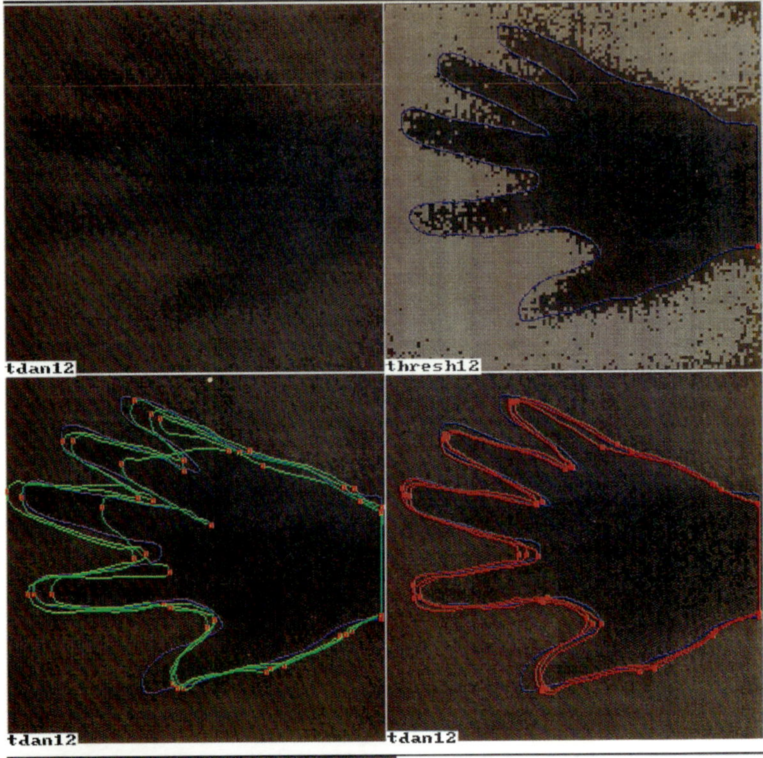

Plate 40

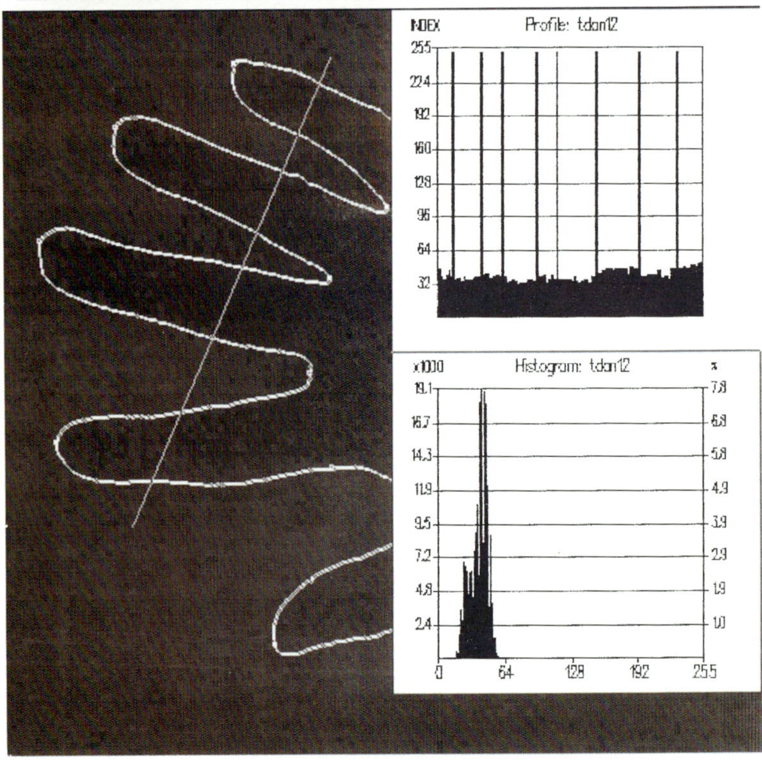

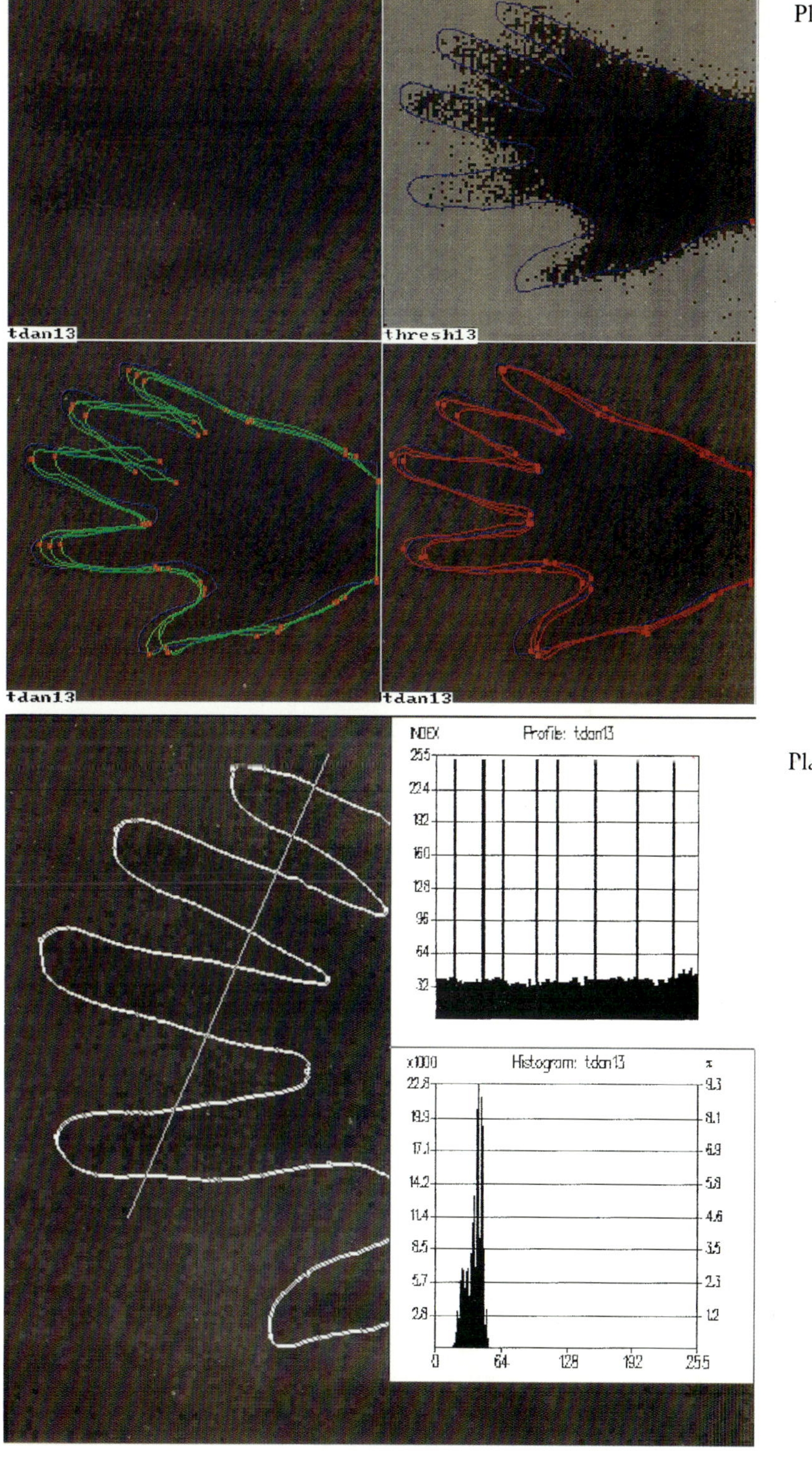

Plate 41

Plate 42

Plate 43

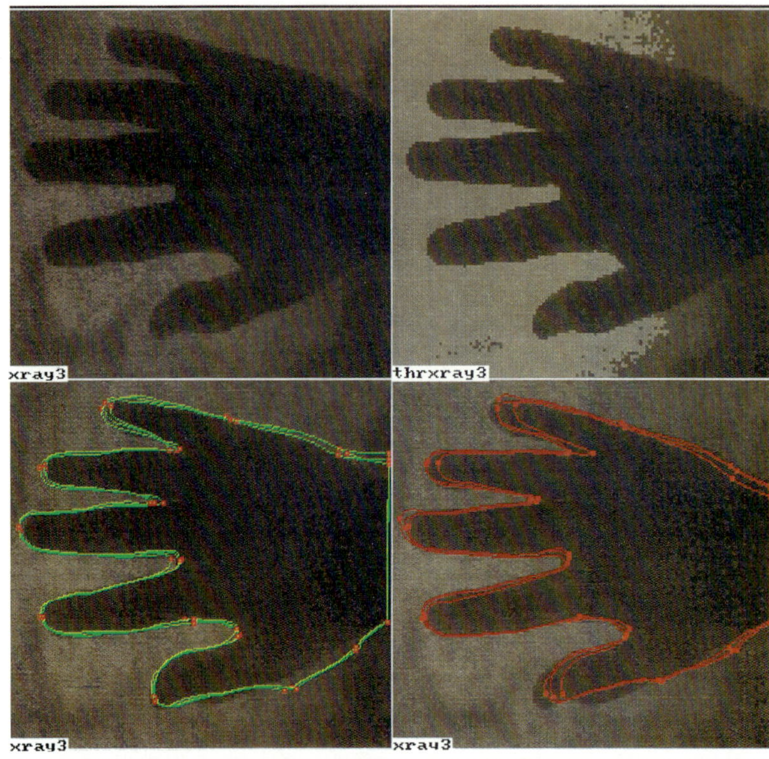

Plate 44

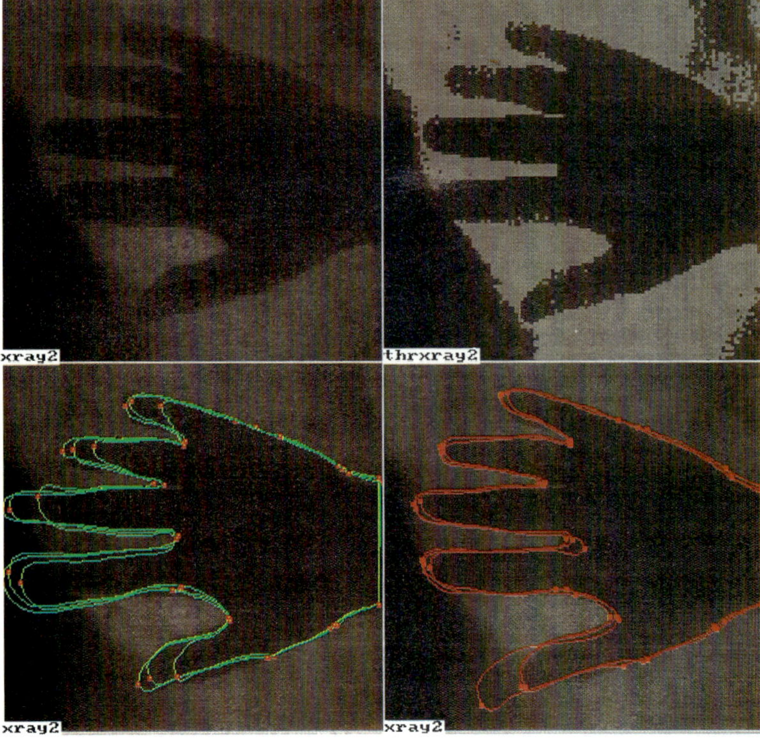

Plate 45

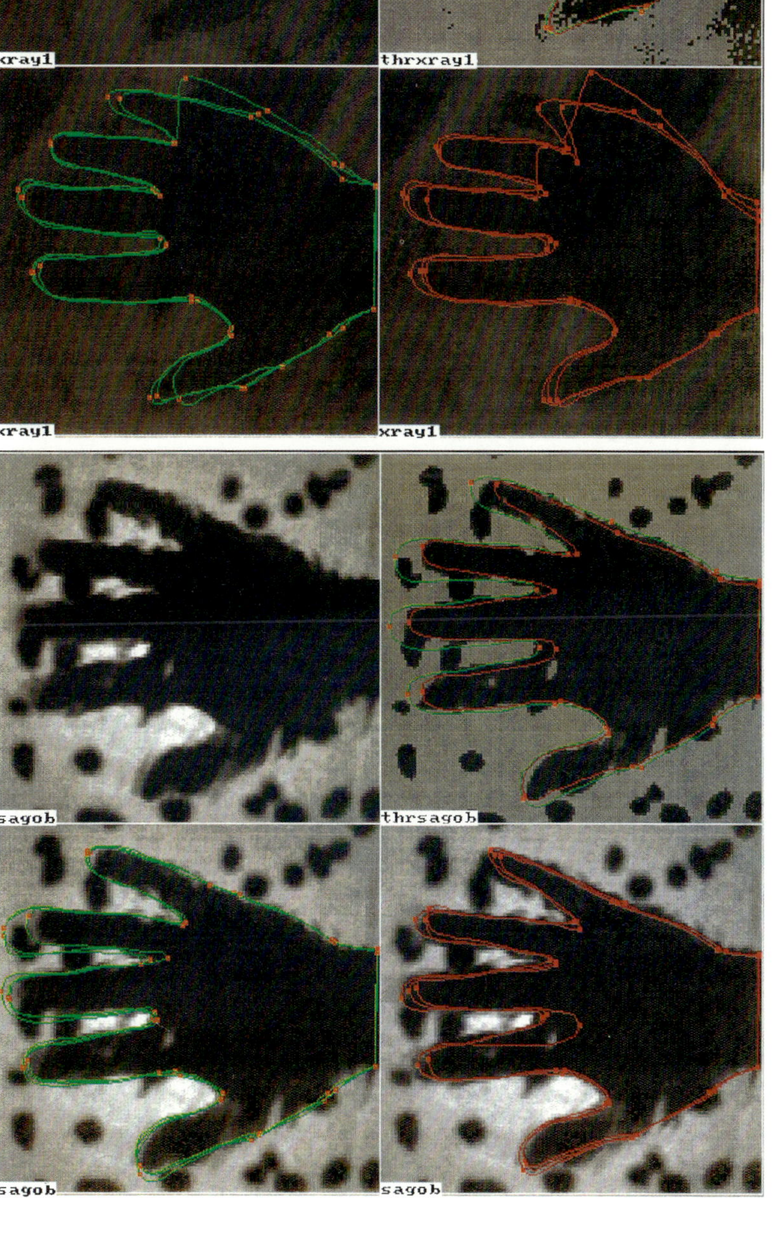

Plate 46

Plate 47

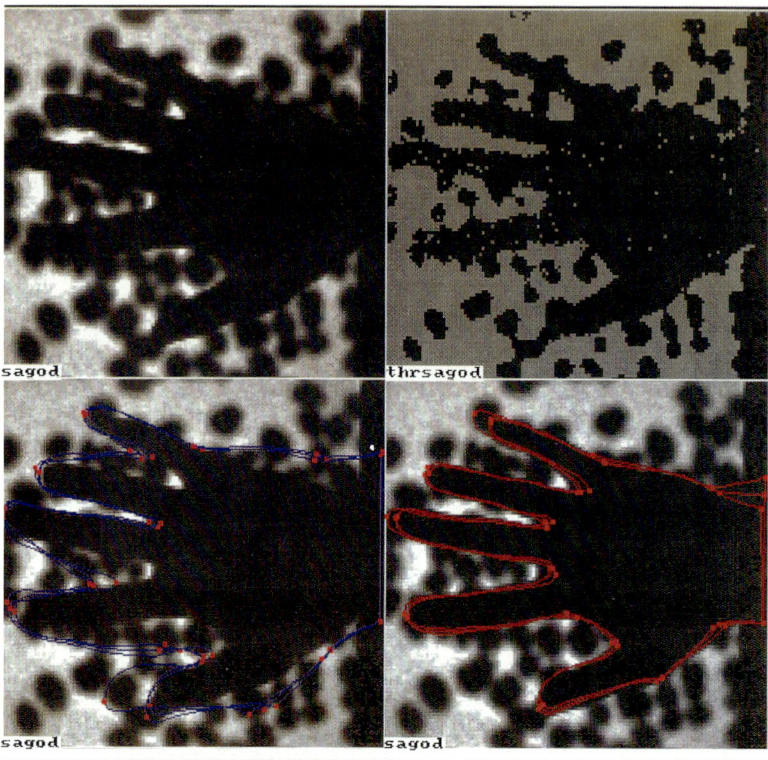

Plate 48

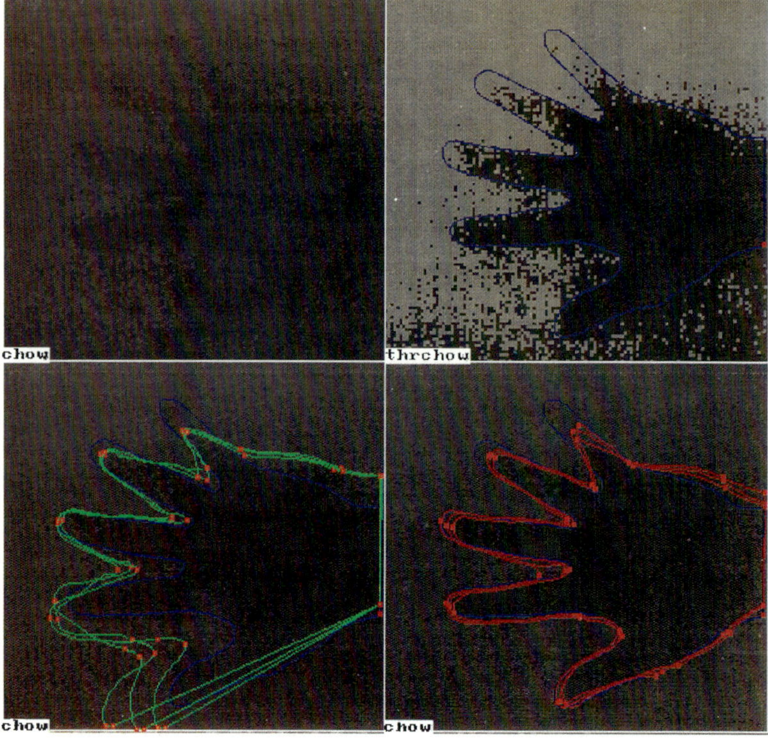

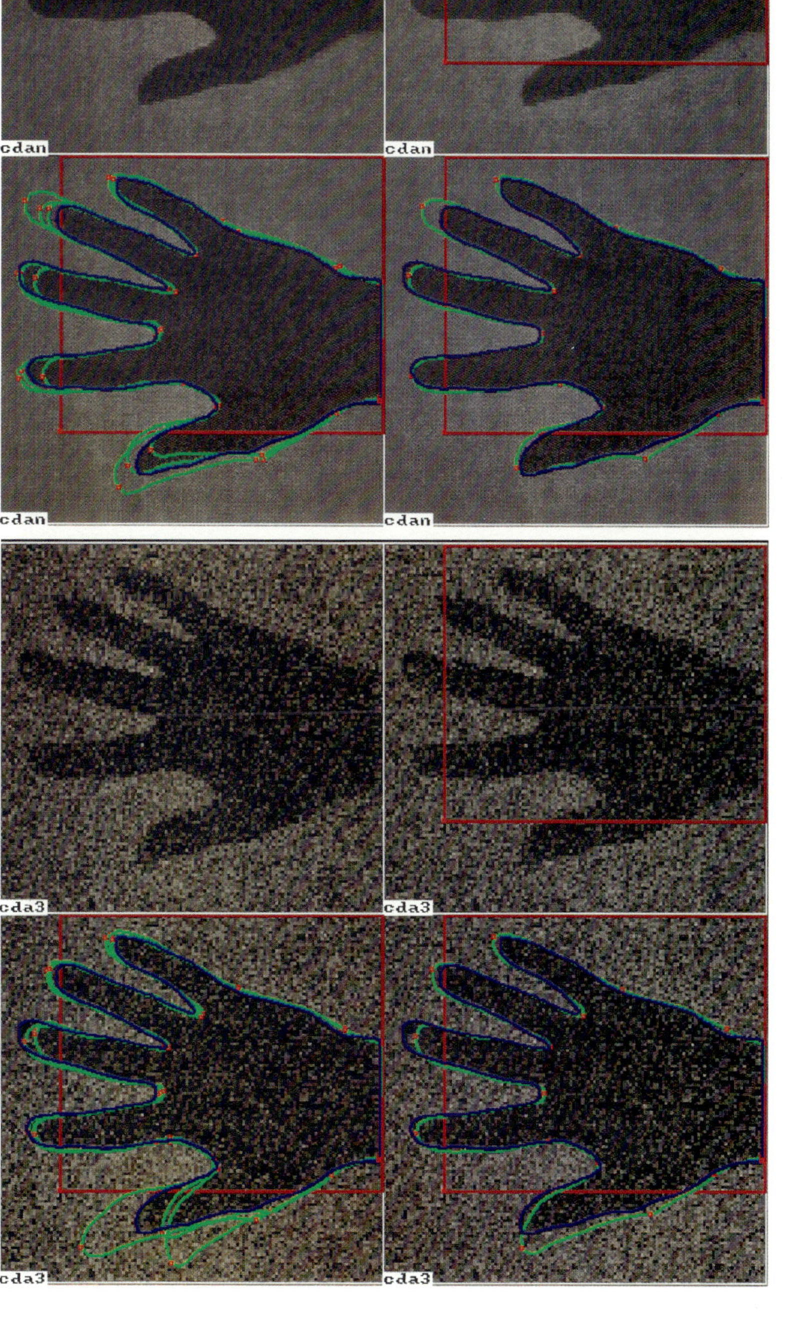

Plate 49

Plate 50

Plate 51

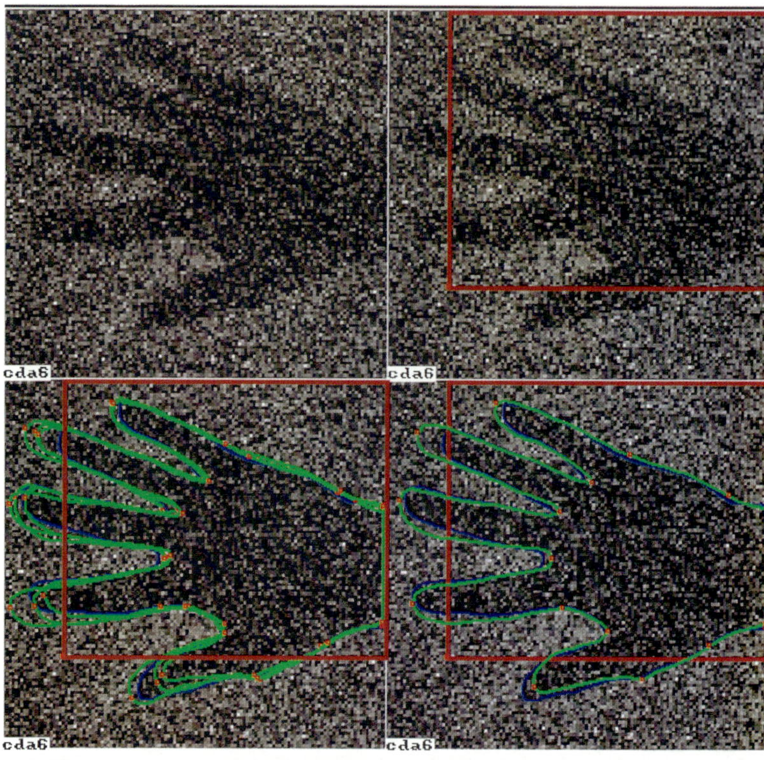

Plate 52

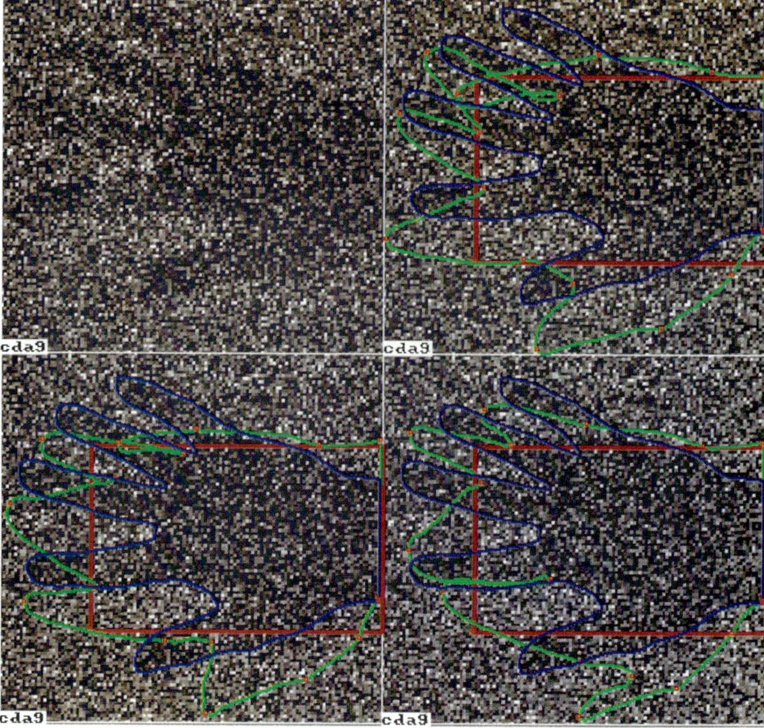

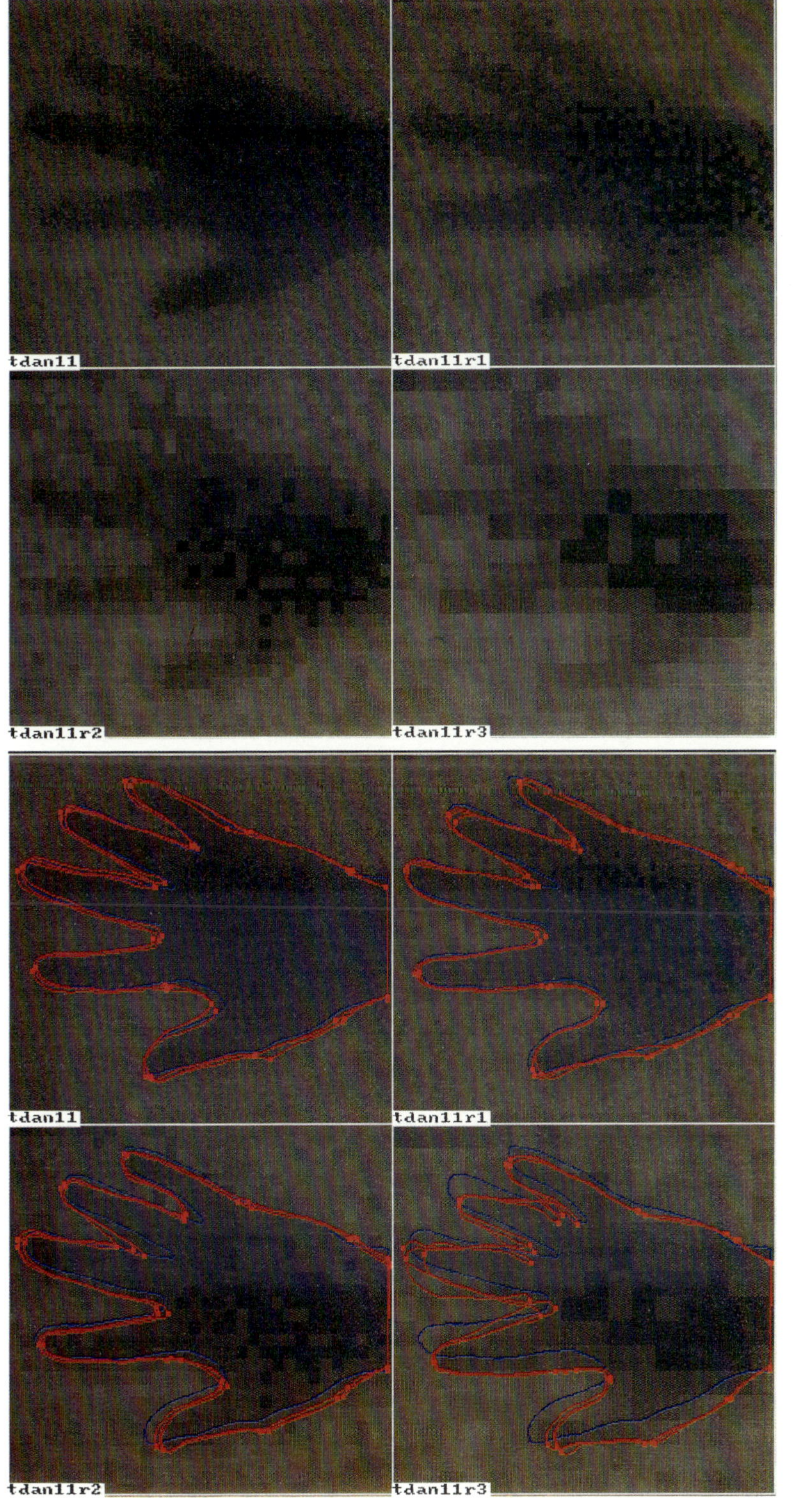

Plate 53

Plate 54

Plate 55

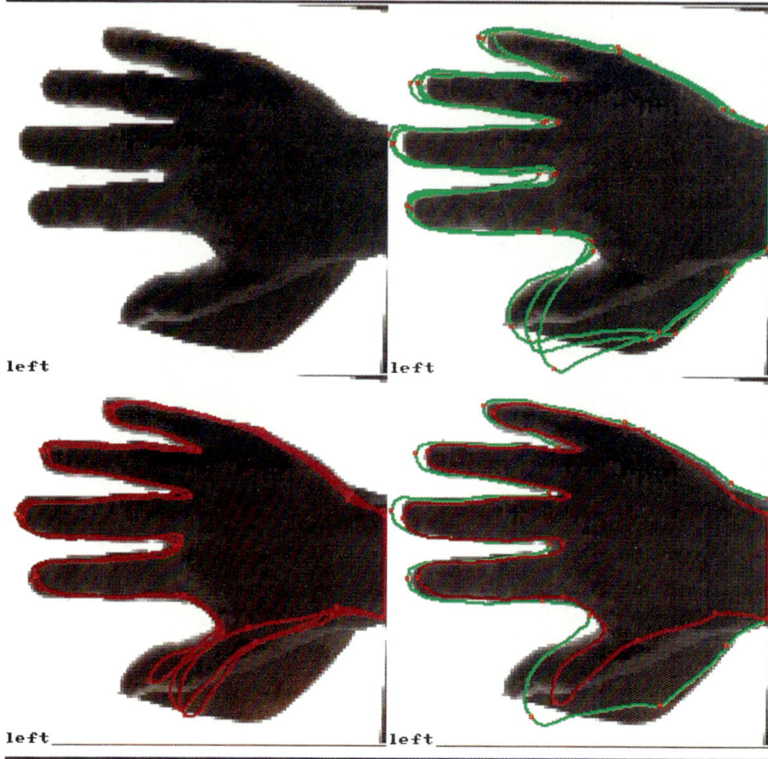

Plate 56

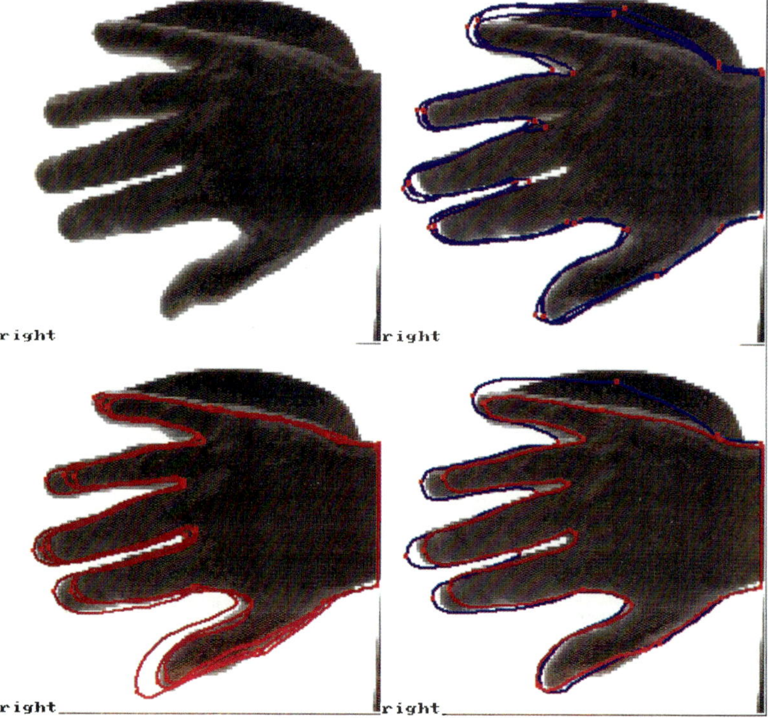

Plate 57

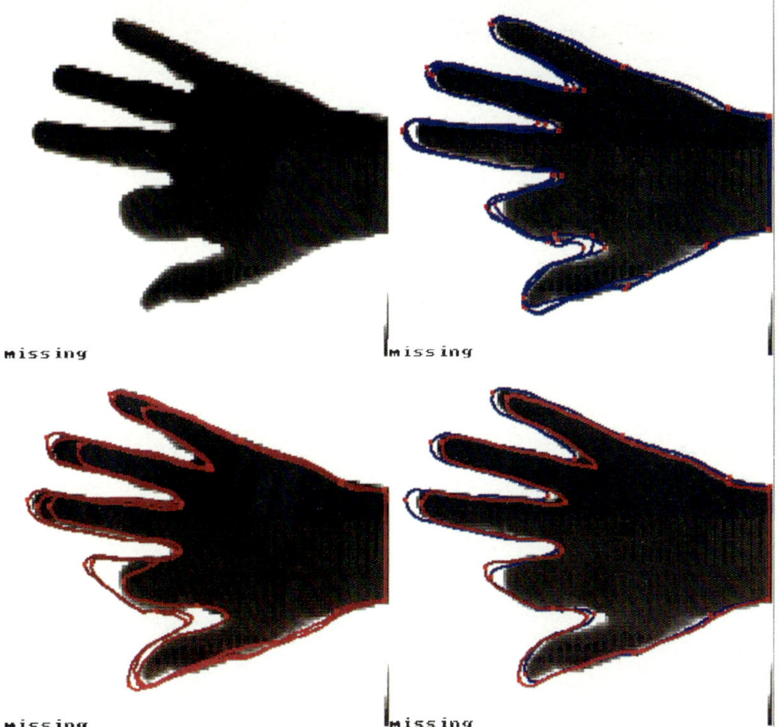

6 Model Critique and Conclusions

§6.1. The experiment has shown that the global shape model has enough *descriptive power* for the analysis of the real pictures we studied. For picture ensembles of the type studied the model will not need any essential modifications except that more attention should be paid to the choice of parameter values, e.g. SIGMAS, SIG2, ABOND.

§6.2. The same is not true for the construction of the deformation mechanism $\mathcal{D}$: it will have to be modified and *tailored to the particular sensor technology* used to capture the pictures. The results were surprisingly robust in terms of $\mathcal{D}$, and were sometimes satisfactory even when the light conditions and optical degradations were substantially different from the $\mathcal{D}$ assumed in the derivation of the algorithms. Nevertheless, we believe that *there is no universally acceptable $\mathcal{D}$; it must be chosen to mirror the acquisition technology.*

§6.3. *The computational feasibility* of the algorithms for the given computational resources was achieved by analytical means: a limit theorem in metric pattern theory, Theorem 2.2, and factored sampling, Theorem 3.1.

No attempt was made to speed up the algorithms even further. No doubt this can be done and should be done.

The development of the mathematically sophisticated algorithms in the computer experiment was facilitated by choosing APL as the programming language. This became strikingly obvious when we compare the 4 months used for code development with the 14 months needed for developing corresponding (although more complex) code in FORTRAN for a parallel experiment, RANGE, to be run on our vector machine STAR100.

§6.4. The success of the experiment opens up an exciting vista in *mathematical image processing.* It dealt with set patterns in 2D, but we can now go ahead to contrast patterns (also in 2D), which is being done in a companion experiment XRAYS where much more biological knowledge is built into the model, especially about bone/cartilege structure. The basic pattern theoretic methodology remains the same however. We are also trying to apply the global shape model to range data acquired by laser radar technology, in an experiment named RANGE. The results of these two experiments will be reported elsewhere.

The flexible geometries used in these experiments are a confirmation.

References

Barry J. Anson, (1963), *An Atlas of Human Anatomy. 2nd Ed.*, Saunders, Philadelphia.

J. Besag, (1974), *Spatial interaction and the statistical analysis of lattice systems*, J. Royal Statist. Soc., Series B **36**, 192-328.

J. Besag, (1986), *On the statistical analysis of dirty pictures*, J. Royal Statist. Soc. **Series B,**, 48.

P. Billingsley, (1968), *Convergence of Probability Measures*, John Wiley & Sons, New York.

F.L. Bookstein, (1978), *The Measurement of Biological Shape and Shape Change*, Lecture Notes in Biomathematics 24,, Springer-Verlag.

T.-S. Chiang, Y. Chow and Y.-I. Lee, (1986), *Evaluation of certain functional integrals*, Preprint.

Y. Chow, (1987), *A limit theorem for pattern synthesis in image processing*, unpublished manuscript.

Y.S. Chow and U. Grenander, (1982), *A singular perturbation problem*, J. Integ.Eqns. **9**, 63-73.

P. Davis, (1979), *Circulant Matrices*, John Wiley & Sons, New York.

W. Feller, (1966), *An Introduction to Probability Theory and its Applications, Vol. II*, John Wiley & Sons, New York.

D. Freedman, (1983), *Brownian Motion and Diffusion*, Springer-Verlag, New York.

S. Geman and D. Geman, (1984), *Stochastic relaxation, Gibbs distributions, and the Bayes restoration of images*, IEEE-PAMI **6**, 721-741.

S. Geman and D.E. McClure, (1985), *Bayesian image analysis: An application to single photon emission tomography*, Proc. Amer. Stat. Assoc., Statistical Computing Section..

B. Gidas, (1985), *Nonstationary Markov chains and convergence of the annealing algorithm*, J. Stat. Physics **39**, 73-131.

U. Grenander, (1970), *A unified approach to pattern analysis*, Adv. in Computers **10, Academic Press**, 175-216.

U. Grenander, (1983), *A Tutorial in Pattern Theory*, Div. Appl. Math., Brown University.

U. Grenander, (1981), *Regular Structures, Part III of Lectures in Pattern Theory*, Springer-Verlag.

U. Grenander and D.M. Keenan, (1986), *On the shape of plane images*, Reports in Pattern Analysis, No. 145, Div. Appl. Math., Brown University.

J.M. Hammersley and D.C. Handscomb, (1964), *Monte Carlo Methods*, Methuen, London.

D.G. Kendall, (1977), *The diffusion of shape*, Adv. in Appl. Prob. **9**, 428-430.

A. Knoerr, (1988), *Global models of natural boundaries: Theory and applications*, Pattern Analysis Tech. Report No. 148, Brown University, Providence, R.I..

K. Knopp, (1945), *Theory of Functions. Part I. 1st American Edition*, Dover Publications, New York.

N.J. Lennes, (1911), *Theorems on the simple finite polygon and polyhedron*, American J. Math. **33**, 37-62.

D. Mitrinović and J. Kečkić, (1984), *The Cauchy Method of Residues*, D. Reidel Publishing Co., Holland.

B. Osborn, (1986), *Parameter estimation in pattern theory, Ph.D. Thesis*, Div. Appl. Math., Brown University.

D.J. Paddon and H. Holstein, eds., (1985), *Multigrid Methods for Integral and Differential Equations*, Clarendon Press, Oxford.

V. Petrov, (1975), *Sums of Independent Random Variables*, Springer-Verlag, New York.

F.P. Preparata and M.I. Shamos, (1985), *Computational Geometry, An introduction*, Springer-Verlag.

D. Revuz, (1984), *Markov Chains. Second Edition*, North-Holland, New York.

B.D. Ripley, (1986), *Statistics, images, and pattern recognition*, Canadian J. of Statist. **14**, 83-111.

J.T. Schwartz and M. Sharir, (1983), *On the "Piano Movers" problem. II. General techniques for computing topological properties of real algebraic manifolds*, Advances in Applied Mathematics **4**, 298-351.

R.L. Tweedie, (1975), *Sufficient conditions for ergodicity and recurrence of Markov chains on a general state space*, Stochastic Processes and their Applications **3**, 385-403.

M. Wichura, (1968), *On the weak convergence of non-Borel probabilities on a metric space*, Ph.D. Thesis, Columbia University.

Appendix A
Experimental Software

The following code is the same as the one used for the experiments except for cosmetic changes. Utilities, as well as plotting and display programs, have not been listed since they depend upon what devices are available.

No systematic attempt has been made to speed up the code except for the analysis parts of the GLOBAL3 and 4 functions; the latter are fast since they can synthesize the whole or almost the whole boundary in one step. We believe that the APL code for restoration could be made much faster. If this is attempted one should begin with the function DEFPROB which computes integrals by expressing them as contour integrals. This can no doubt be improved considerably.

We have left fairly many variables global for easy access. To save space they should be made global and erased from the workspace.

To facilitate using or modifying we give the meaning of some of the global variables used.

name	meaning
ABOND	$2 \times N$ matrix of coupling constants
ACCEPT	Boolean vector representing areas with MIN less than MOUT
ACHORD	vector defined in section 3
CHORD	vector defined in section 3
HOR	numeric vector of horizontal coordinates of trial area
ILEFT	left site fixed in sweep area of connector graph
IRIGHT	right site fixed in sweep area of connector graph
KEEP	Boolean SS-vector representing sample subscripts corresponding to arcs inside rectangle
LARGE	Boolean variable to be put = 1 if trial area is very large, = 0 else
LH	horizontal dimension of digital picture
LHOR	horizontal dimension trial area
LV	vertical dimension of digital picture
LVERT	vertical dimension of trial area
MASKH	horizontal dimension of area left after masking
MASKIN	Boolean $LV \times LH$ matrix describing mask for estimating MIN
MASKOUT	Boolean $LV \times LH$ matrix describing mask for estimating MOUT
MASKV	vertical dimension of area left after masking
MIN	mean intensity – inside hand
MOUT	mean intensity – outside hand
N	number of vertices of current template
NEWLAND	vector of hinges as indexed in TEMP256
NEWVERTICES	3-dimensional array describing the generated arcs from ILEFT to IRIGHT
NIN	number of lattice points inside integration region

name	meaning
NKEEP	number of elements in vector KEEP
NOUT	number of lattice points outside integration region
RANDOM	Boolean variable that should be made = 1 for randomization, = 0 else
RHO	average coupling coefficient
RHOS	vector of coupling coefficients
S	array containing group elements
SEEN	Boolean $LV \times LH$ matrix describing area not masked out
SEENH	horizontal dimension of SEEN
SEENV	vector of σ-values
SIG2	$2 \times N$ array of σ-values, first row for $\sigma_{00} = s_{11}$, second row for $\sigma_{10} = -\sigma_{01}$
SITES	2-column matrix of site values describing sweep strategy, each row has left and right endpoint for one sweep area
SS	sample size for trials in each sweep area, see sections 3 and 4
SVALUES	array of trial generated group elements
SYNTHESIS	Boolean variable that should be put = 1 for simulation, = 0 else
TAV	variances of intensity values
TEMPLATE	current template
TEMPLATE1	fixed initial template with $N = 36$ or 46
TEMPLATE256	full template with 256 vertices
VERTICES	vertices selected after random generation
VIN	variance estimated within integration area
VOUT	variance estimated outside integration area

```
      ∇ VRTSD←ILEFT VERTEXSD IRIGHT
       ⎕IO←0
       VERTICES←TEMPLATE
       P←1+N|IRIGHT-ILEFT
       VRTSD←(3,P-1)⍴0
       CHORD←VERTICES[IRIGHT;]-VERTICES[ILEFT;]
       LEFTS←RIGHTS←I2← 2 2 ⍴ 1 0 0 1
       IDEN←(⍳P-1)∘.=(⍳P-1)
       IL1←ILEFT+1+⍳P-1
       IL2←ILEFT+1+⍳P-2
       L←(2,(P-1),(P-1))⍴0
       KC←0
      LOOP1:A1←(÷SIG2[KC;N|IL1-1])+(ABOND[KC;N|IL1]*2)÷SIG2[KC;N|IL1]
       A2←-ABOND[KC;N|IL2]÷SIG2[KC;N|IL2]
       A3←(A2∘.×(P-2)⍴1)×IDEN[⍳P-2;⍳P-2]
       A4←((P-1)⍴0),[1] A3,[0]((P-2)⍴0)
       A5←(((P-2)⍴0),[0] A3),[1](P-1)⍴0
       M←((A1∘.×(P-1)⍴1)×IDEN)+A4+A5
       NM←⌹M
       ⎕IO←1
       LA←CHOLESKYR(CONVERT NM)
       L[KC+1;;]←LA[REVERSE1 P-1]×((⍳P-1)∘.≥(⍳P-1))
       ⎕IO←0
       KC←KC+1
       →(KC≤1)/LOOP1
       A2←GA←⍉TEMPLATE[N|IL1;]-TEMPLATE[N|IL1-1;]
       A2[0;]←-GA[0;]
       A←(2,(2×P-1))⍴(GA,[0](⊖A2))
       CONS←((P-1)⍴1),((P-1)⍴0)
       ZER←((P-1),(P-1))⍴0
       U1←(L[0;;],[1] ZER),[0](ZER,[1] L[1;;])
       AL←A+.×U1
       V←AL+.×⍉AL
       VINV←⌹V
       W←U1+.×(⍉AL)+.×VINV
       PROJ←((⍳2×P-1)∘.=(⍳2×P-1))-((⍉AL)+.×VINV+.×AL)
       LT←U1+.×PROJ
       MU←(2×P-1)⍴0
       AS1←ABOND[0;ILEFT]÷SIG2[0;ILEFT]
       AS2←ABOND[0;IRIGHT]÷SIG2[0;IRIGHT]
       MU[0]←AS1×(LEFTS[0;0]-I2[0;0])
       MU[P-2]←AS2×(RIGHTS[0;0]-I2[0;0])
       AS1←ABOND[1;ILEFT]÷SIG2[1;ILEFT]
       AS2←ABOND[1;IRIGHT]÷SIG2[1;IRIGHT]
       MU[P-1]←AS1×(LEFTS[0;1]-I2[0;1])
       MU[(2×P)-3]←AS2×(RIGHTS[0;1]-I2[0;1])
       MU←U1+.×(⍉U1)+.×MU
       NU←A+.×MU
       Z←MU+W+.×((CHORD-A+.×CONS)-NU)
       REARR←((2×P-1),(2×P-1))⍴,((IDEN,[1] ZER),[1](ZER,[1] IDEN))
       RAR←REARR+.×LT
       CON←RAR+.×⍉RAR
       AB←A+.×⍉REARR
       KDIM←0
```

```
LOOP2:KDIM←KDIM+1
 ATE←AB[0;⍳2×KDIM]
 BTE←AB[1;⍳2×KDIM]
 VRTSD[0;KDIM-1]←ATE+.×CON[⍳2×KDIM;⍳2×KDIM]+.×ATE
 VRTSD[1;KDIM-1]←ATE+.×CON[⍳2×KDIM;⍳2×KDIM]+.×BTE
 VRTSD[2;KDIM-1]←BTE+.×CON[⍳2×KDIM;⍳2×KDIM]+.×BTE
 →(KDIM<P-1)/LOOP2
 VRTSD[0 2 ;]←(|VRTSD[0 2 ;])*0.5
 VRTSD[1;]←VRTSD[1;]÷(VRTSD[0;]×VRTSD[2;])
 ∇
 >
```

```
 ∇ I UPDATE1 J;ARCL;V;I;J;TEMP;TEMP1
⍝FILLS IN INTERMEDIATE VERTICES FROM TEMPLATE2
 ARCL←256|NEWLAND[J]-NEWLAND[I]
 TEMP1←(ARCLρ1)∘.×VERTICES[N|I;]
 V←256|1+NEWLAND[I]+⍳ARCL
⍝VECTOR COARSE OBTAINED FROM FUNCTION ALSO CALLED COARSE
 NVERTICES[V;]←TEMP1++\[0](0 1 0)⍉S[COARSE[256|V-1];;]+.×⍉TEMPLATE2[V;]
 TEMPLATE2[256|V-1;]
 ∇

 ∇ I UPDATE2 J;ARCL;V;I;J;TEMP;TEMP1
⍝FILLS IN INTERMEDIATE VERTICES
 ARCL←256|NEWLAND[J]-NEWLAND[I]
 TEMP←(ARCLρ1)∘.×TEMPLATE[N|I;]
 TEMP1←(ARCLρ1)∘.×VERTICES[N|I;]
 V←256|NEWLAND[I]+⍳ARCL
 NVERTICES[V;]←TEMP1+⍉S[N|I;;]+.×⍉TEMPLATE2[V;]-TEMP
 NS[V;;]←(ARCLρ1)∘.×S[N|I;;]
 ∇
```

```
  ∇ SVALUES←RANDOMGROUP1 SS;I2;IDEN;A1;A2;A3;A4;A5;M;NM;⎕IO;L;A;CONS;U1;AL;V
    ;W;PROJ;LT;MU;NU;NOR;Z;ZZ;IL1;IL2;AS1;AS2;ROW
  ⍝GENERATES RANDOM ELEMENTS FROM GROUP    GL(2) ALONG P-1 ARCS OF
  ⍝THE TEMPLATE.  GIVEN OUTSIDE THE PROGRAM ARE P,ILEFT,IRIGHT,
  ⍝LEFTS,RIGHTS,AND THE N-ARRAY PARAMETERS SIG2 AND ABOND.
  ⎕IO←0
  I2← 2 2 ⍴ 1 0 0 1
  SVALUES←(SS,(P-1),2,2)⍴I2
  IDEN←(⍳P-1)∘.=(⍳P-1)
  IL1←ILEFT+1+⍳P-1
  IL2←ILEFT+1+⍳P-2
  A1←(÷SIG2[N|IL1-1])+(ABOND[N|IL1]*2)÷SIG2[N|IL1]
  A2←-ABOND[N|IL2]÷SIG2[N|IL2]
  A3←(A2∘.×(P-2)⍴1)×IDEN[⍳P-2;⍳P-2]
  A4←((P-1)⍴0),[1] A3,[0]((P-2)⍴0)
  A5←(((P-2)⍴0),[0] A3),[1](P-1)⍴0
  M←((A1∘.×(P-1)⍴1)×IDEN)+A4+A5
  NM←⌹M
  ⎕IO←1
  L←CHOLESKYR(CONVERT NM)
  L←L[REVERSE1 P-1]×((⍳P-1)∘.≥(⍳P-1))
  ⎕IO←0
  A←,⍉TEMPLATE[N|IL1;]-TEMPLATE[N|IL1-1;]
  CONS←(((P-1)⍴1)∘.×(1 0)),[0]((P-1)⍴1)∘.×(0 1)
  U1←((2×P-1),(2×P-1))⍴((0 2 1 3)⍉(I2∘.×L))
  AL←A+.×U1
  V←AL+.×⍉AL
  W←(÷V)×U1+.×⍉AL
  PROJ←((⍳2×P-1)∘.=(⍳2×P-1))-((AL∘.×AL)÷V)
  LT←U1+.×PROJ
  ROW←0
 LOOP:MU←(2×P-1)⍴0
  AS1←ABOND[ILEFT]÷SIG2[ILEFT]
  AS2←ABOND[IRIGHT]÷SIG2[IRIGHT]
  MU[0]←AS1×(LEFTS[ROW;0]-I2[ROW;0])
  MU[P-2]←AS2×(RIGHTS[ROW;0]-I2[ROW;0])
  MU[P-1]←AS1×(LEFTS[ROW;1]-I2[ROW;1])
  MU[(2×P)-3]←AS2×(RIGHTS[ROW;1]-I2[ROW;1])
  MU←U1+.×(⍉U1)+.×MU
  NU←+/A×MU
  NOR←¯6+1E¯8×+/?(SS,(2×P-1),12)⍴100000000
  Z←MU+W×((CHORD[ROW]-+/A×CONS[;ROW])-NU)
  ZZ←(Z∘.×(SS⍴1))+LT+.×⍉NOR
  SVALUES[;;ROW;]←SVALUES[;;ROW;]+(2 1 0)⍉((2,(P-1),SS)⍴ZZ)
  ROW←ROW+1
  →(ROW<2)/LOOP
  ∇
```

```
     ∇ SVALUES←RANDOMGROUP2 SS;IDENTITY;M0;M1;US
[1]   ⍝SIMULATES  SAMPLE OF SIZE SS OF GAUSSIAN MEASURE
[2]   ⍝ON GROUP  GL(2)  WITH PARAMETERS SIGMA AND RHOLEFT,RHORIGHT
[3]   ⍝INDEX ORIGIN=0
[4]   ⍝LEFT VALUE OF CURRENT S-VALUE IS CALLED LEFT,
[5]   ⍝THE RIGHT ONE IS CALLED RIGHT
[6]   ⍝NON-ZERO 2-VECTOR CALLED ACHORD IS TRANSFORMED BY RANDOM GROUP E
      LEMENT
[7]   ⍝INTO GLOBAL VECTOR CHORD
[8]    A2←+/ACHORD*2
[9]    SVALUES←(SS,2,2)ρ0
[10]   IDENTITY← 2 2 ρ 1 0 0 1
[11]   TLEFT←LEFTS-IDENTITY
[12]   TRIGHT←RIGHTS-IDENTITY
[13]   M0←0.5×(RHOLEFT×TLEFT[0;0])+RHORIGHT×TRIGHT[0;0]
[14]   M1←0.5×(RHOLEFT×TLEFT[0;1])+RHORIGHT×TRIGHT[0;1]
[15]   D←CHORD[0]-ACHORD[0]
[16]   U←¯6+1E¯8×+/?(SS,12)ρ100000000
[17]   U←U×SIGMA÷A2*0.5
[18]   U0←(ACHORD[1]×U)+((M0×ACHORD[1]*2)+(ACHORD[0]×D)-ACHORD[0]×
       ACHORD[1]×M1)÷A2
[19]   U1←(-ACHORD[0]×U)+((-M0×ACHORD[0]×ACHORD[1])+(ACHORD[1]×D)+M1×
       ACHORD[0]*2)÷A2
[20]   SVALUES[;0;0]←1+U0
[21]   SVALUES[;0;1]←U1
[22]   M0←0.5×(RHOLEFT×TLEFT[1;0])+RHORIGHT×TRIGHT[1;0]
[23]   M1←0.5×(RHOLEFT×TLEFT[1;1])+RHORIGHT×TRIGHT[1;1]
[24]   D←CHORD[1]-ACHORD[1]
[25]   U←¯6+1E¯8×+/?(SS,12)ρ100000000
[26]   U←U×SIGMA÷A2*0.5
[27]   U0←(ACHORD[1]×U)+((M0×ACHORD[1]*2)+(ACHORD[0]×D)-ACHORD[0]×
       ACHORD[1]×M1)÷A2
[28]   U1←(-ACHORD[0]×U)+((-M0×ACHORD[0]×ACHORD[1])+(ACHORD[1]×D)+M1×
       ACHORD[0]*2)÷A2
[29]   SVALUES[;1;0]←U0
[30]   SVALUES[;1;1]←U1+1
     ∇
```

```
∇ SVALUES←RANDOMGROUP3 SS;I2;IDEN;A1;A2;A3;A4;A5;M;NM;LA;⎕IO;L;A
   ;CONS;ZER;U1;AL;V;W;PROJ;LT;MU;NU;NOR;Z;ZZ;IL1;IL2;AS1;AS2;KC
⍝GENERATES SS SAMPLES FROM GAUSSIAN MEASURE ON GROUP  O(2) WHERE
⍝THE PARAMETERS SIG2 AND ABOND ARE 2×N ARRAYS AS OPPOSED TO
⍝THE SITUATION OF RANDOMGROUP1 WHERE THEY ARE N-VECTORS.
 ⎕IO←0
 I2← 2 2 ρ 1 0 0 1
 SVALUES←(SS,(P-1),2,2)ρI2
 IDEN←(⍳P-1)∘.=(⍳P-1)
 IL1←ILEFT+1+⍳P-1
 IL2←ILEFT+1+⍳P-2
 L←(2,(P-1),(P-1))ρ0
 KC←0
LOOP1:A1←(÷SIG2[KC;N|IL1-1])+(ABOND[KC;N|IL1]*2)÷SIG2[KC;N|IL1]
 A2←-ABOND[KC;N|IL2]÷SIG2[KC;N|IL2]
 A3←(A2∘.×(P-2)ρ1)×IDEN[⍳P-2;⍳P-2]
 A4←((P-1)ρ0),[1] A3,[0]((P-2)ρ0)
 A5←(((P-2)ρ0),[0] A3),[1](P-1)ρ0
 M←((A1∘.×(P-1)ρ1)×IDEN)+A4+A5
 NM←⌹M
 ⎕IO←1
 LA←CHOLESKYR(CONVERT NM)
 L[KC+1;;]←LA[REVERSE1 P-1]×((⍳P-1)∘.≥(⍳P-1))
 ⎕IO←0
 KC←KC+1
 →(KC≤1)/LOOP1
 A2←GA←⍉TEMPLATE[N|IL1;]-TEMPLATE[N|IL1-1;]
 A2[0;]←-GA[0;]
 A←(2,(2×P-1))ρ(GA,[0](⊖A2))
 CONS←((P-1)ρ1),((P-1)ρ0)
 ZER←((P-1),(P-1))ρ0
 U1←(L[0;;],[1] ZER),[0](ZER,[1] L[1;;])
 AL←A+.×U1
 V←AL+.×⍉AL
 VINV←⌹V
 W←U1+.×(⍉AL)+.×VINV
 PROJ←((⍳2×P-1)∘.=(⍳2×P-1))-((⍉AL)+.×VINV+.×AL)
 LT←U1+.×PROJ
 MU←(2×P-1)ρ0
 AS1←ABOND[0;ILEFT]÷SIG2[0;ILEFT]
 AS2←ABOND[0;IRIGHT]÷SIG2[0;IRIGHT]
 MU[0]←AS1×(LEFTS[0;0]-I2[0;0])
 MU[P-2]←AS2×(RIGHTS[0;0]-I2[0;0])
 AS1←ABOND[1;ILEFT]÷SIG2[1;ILEFT]
 AS2←ABOND[1;IRIGHT]÷SIG2[1;IRIGHT]
 MU[P-1]←AS1×(LEFTS[0;1]-I2[0;1])
 MU[(2×P)-3]←AS2×(RIGHTS[0;1]-I2[0;1])
 MU←U1+.×(⍉U1)+.×MU
 NU←A+.×MU
 NOR←¯6+1E¯8×+/?(SS,(2×P-1),12)ρ100000000
 Z←MU+W+.×((CHORD-A+.×CONS)-NU)
 ZZ←(Z∘.×(SSρ1))+LT+.×⍉NOR
 SVALUES[;;0;]←SVALUES[;;0;]+(2 1 0)⍉((2,(P-1),SS)ρZZ)
 SVALUES[;;1;]←(⊖[2] SVALUES[;;0;])×((SSρ1)∘.×((P-1)ρ1)∘.×(¯1 1))
∇
```

```
    ∇ NGON;V;XS;YS
    ⍝COMPUTES VERTICES C IN  REGULAR       N-GON
     V←○2×(0.5+⍳N)÷N
     XS←0.5+RADIUS×2○V
     YS←0.5+RADIUS×1○V
     C←⍉(2,N)⍴XS,YS
    ∇
```

```
    ∇ L←A LTR1 X;N;I
    ⍝ CALCULATES ELEMENTS BELOW THE MAIN DIAGONAL FOR CHOLESKYR
    ⍝ LAST UPDATE: 6/19/79
     X←,X
     A←,A
    S1:N←(¯1+((1+8×⍴A)*0.5))÷2
     I←1
     L←,X[1]÷A[1]
    S2:→0×⍳N<I←I+1
     L←L,(÷A[J+I])×X[I]-A[(J←I×J÷2)+⍳J]+.×(J←I-1)↑L
     →S2
    ∇
```

```
    ∇ LOCALEST VS1;VS2;BOOLEAN;COND;TEMP
    ⍝PREPARES FOR  LOCAL ESTIMATION OF  PARAMETERS     IN DEFORMATION
      MECHANISM
    ⍝CALLED BY GLOBAL4
    ⍝NEEDS GLOBAL VARIABLES HOR AND VERT
    ⍝INDEX ORIGIN 0
     LVERT1←LVERT
     LHOR1←LHOR
     VERT1←VERT
     HOR1←HOR
     VERT←⍳LVERT←LV
     HOR←⍳LHOR←LH
     VS2←1⊖[0] VS1
     BOOLEAN←(((VS1[;0]⌈VS2[;0])∘.≥VERT)∧((VS1[;0]⌊VS2[;0])∘.≤VERT))∘
     .∧LHOR⍴1
     TEMP←VS2[;0]-VS1[;0]
     COND←(VS1[;1]∘.×LVERT⍴1)+(VS1[;0]∘.-VERT)×((VS1[;1]-VS2[;1])÷
     TEMP+0=TEMP)∘.×LVERT⍴1
     COND←COND∘.≥HOR
     BOOLEAN←BOOLEAN∧COND
     POS←+/[0]((TEMP<0)∘.∧(LVERT⍴1)∘.∧(LHOR⍴1))∧BOOLEAN
     NEG←+/[0]((TEMP>0)∘.∧(LVERT⍴1)∘.∧(LHOR⍴1))∧BOOLEAN
     LVERT←LVERT1
     LHOR←LHOR1
     VERT←VERT1
     HOR←HOR1
     BOOLEAN←(POS-NEG)[VERT;HOR]
     SUMT1←+/+/BOOLEAN
     SUMTX←+/+/BOOLEAN×ID[VERT;HOR]
     SUMTX2←+/+/BOOLEAN×ID[VERT;HOR]*2
    ∇
```

```
      ∇ LARGEF;V1;T2
    ⍝MODIFIES RESTORATION ALGORITH FOR GLOBAL1
    ⍝IN THE CASE OF LARGE VALUES FOR LVERT AND LHOR
    ⍝TO REDUCE STORAGE REQUIREMENT
     T2←0
     SUMX←SUM1←0
    LOOP1:V1←⍉NEWVERTICES[K;;242|START[T2]+⍳LENGTH[T2]]
     V1←V1,[0] 64.001 60.001
     SUMX←SUMX+DEFPROB V1
     SUBIMAGE←(LVERT,LHOR)⍴1
     SUM1←SUM1+DEFPROB V1
     T2←T2+1
     →(T2<4)/LOOP1
      ∇
```

```
      ∇ HANDSYNTH1;NVERTICES;NS
    ⍝CARRIES OUT PATTERN SYNTHESIS  USING GLOBAL1
    ⍝AND RANDOMGROUP1
    ⍝      POLYGON      TEMPLATE SHOULD HAVE N VERTICES
    ⍝ BEING A COARSE VERSION OF TEMPLATE256 WITH 256 VERTICES
     S←(N⍴1)∘.× 2 2 ⍴ 1 0 0 1
     NS←(256⍴1)∘.× 2 2 ⍴ 1 0 0 1
     RANDOM←SYNTHESIS←1
     SS←2
     VERTICES←TEMPLATE
     NVERTICES←TEMPLATE2←TEMPLATE256
     1 GLOBAL1 0
     1 UPDATE1 0
     →(NKEEP<2)/WARNING
     VERTICES←NVERTICES
     →0
    WARNING:'RESULT OUTSIDE RECTANGLE'
      ∇
```

```
    ∇ HANDSYNTH3;NVERTICES;NS
  ⍝CARRIES OUT PATTERN SYNTHESIS USING GLOBAL3
  ⍝AND RANDOMGROUP3
  ⍝      POLYGON      TEMPLATE SHOULD HAVE N VERTICES
  ⍝ BEING A COARSE VERSION OF TEMPLATE256 WITH 256 VERTICES
   S←(N⍴1)∘.× 2 2 ⍴ 1 0 0 1
   NS←(256⍴1)∘.× 2 2 ⍴ 1 0 0 1
   RANDOM←SYNTHESIS←1
   SS←2
   VERTICES←TEMPLATE
   NVERTICES←TEMPLATE2←TEMPLATE256
   1 GLOBAL3 0
   1 UPDATE1 0
   →(NKEEP<2)/WARNING
   VERTICES←NVERTICES
   →0
  WARNING:'RESULT OUTSIDE RECTANGLE'
    ∇

    ∇ HANDSYNTH23
  ⍝SYNTHESIZES HAND SHAPES WITH GLOBAL
  ⍝               MODEL USING FIRST GLOBAL2 AND THEN GLOBAL3
  ⍝SET PARAMETERS OUTSIDE PROGRAM
   SS←2
   RANDOM←1
   SYNTHESIS←1
   LARGE←0
   VERTICES←TEMPLATE←TEMPLATE1
   S←(N⍴1)∘.× 2 2 ⍴ 1 0 0 1
   NS←(256⍴1)∘.× 2 2 ⍴ 1 0 0 1
   1 GLOBAL2 0
   1 UPDATE2 0
   →(NKEEP<2)/WARNING
   TEMPLATE←VERTICES
   TEMPLATE2←NVERTICES
   S←(N⍴1)∘.× 2 2 ⍴ 1 0 0 1
   1 GLOBAL3 0
   →(NKEEP<2)/WARNING
   1 UPDATE1 0
   VERTICES←NVERTICES
   →0
  WARNING:'RESULT OUTSIDE OF RECTANGLE'
    ∇
```

```
    ∇ HANDRESTS ITER;T
 ⍝CARRIES OUT IMAGE RESTORATION OF DEFORMED PICTURE ID
 ⍝CALLS GLOBAL4   AND UPDATE2 AS WELL AS GLOBAL2 AND UPDATE1
 ⍝CARRIES OUT IMAGE EXTRAPOLATION
 ⍝ALLOWS PART OF PICTURE TO BE MASKED
  VERTICES←TEMPLATE←TEMPLATE1
  S←(N⍴1)∘.× 2 2 ⍴ 1 0 0 1
  NVERTICES←TEMPLATE2←TEMPLATE256
  NS←(256⍴1)∘.× 2 2 ⍴ 1 0 0 1
  SIGMAS←256⍴0.3
  ESTIMATEM
  1 GLOBAL2M 0
  1 UPDATE2 0
  TEMPLATE←VERTICES
  TEMPLATE2←NVERTICES
  S←(N⍴1)∘.× 2 2 ⍴ 1 0 0 1
  T←0
  T1←0
  SIG2←(2,N)⍴0.02
  ABOND←(2,N)⍴0.8
 LOOP0:T←(⍴SITES)[0]|T
  I←SITES[T;0]
  J←SITES[T;1]
  I GLOBAL5 J
  I UPDATE1 J
  T←T+1
  T1←T1+1
  →(T1≤ITER×(⍴SITES)[0])/LOOP0
  TEMPLATE←VERTICES
  TEMPLATE2←NVERTICES
  S←(N⍴1)∘.× 2 2 ⍴ 1 0 0 1
  T←0
  T1←0
  SIG2←(2,N)⍴0.01
  ABOND←(2,N)⍴ 0.6 5
 LOOP:T←(⍴SITES)[0]|T
  I←SITES[T;0]
  J←SITES[T;1]
  I GLOBAL5 J
  I UPDATE1 J
  T←T+1
  T1←T1+1
  →(T1≤ITER×(⍴SITES)[0])/LOOP
  VERTICES←NVERTICES
    ∇
```

```
      ∇ HANDREST4 ITER;T
   ⍝CARRIES OUT IMAGE RESTORATION OF DEFORMED PICTURE ID
   ⍝CALLS GLOBAL4   AND UPDATE2 IN ADDITION TO GLOBAL2
   ⍝USES LOCAL ESTIMATION
   ⍝AND 2- OR 3-STAGE MODEL
    VERTICES←TEMPLATE←TEMPLATE1
    S←(Nρ1)∘.× 2 2 ρ 1 0 0 1
    NVERTICES←TEMPLATE2←TEMPLATE256
    NS←(256ρ1)∘.× 2 2 ρ 1 0 0 1
    SIGMAS←256ρ0.3
    ESTIMATE
    1 GLOBAL2 0
    1 UPDATE2 0
    TEMPLATE←VERTICES
    TEMPLATE2←NVERTICES
    S←(Nρ1)∘.× 2 2 ρ 1 0 0 1
    T←0
    T1←0
    SIG2←(2,N)ρ0.02
    ABOND←(2,N)ρ0.8
   LOOP0:T←(ρSITES)[0]|T
    I←SITES[T;0]
    J←SITES[T;1]
    I GLOBAL3 J
    I UPDATE1 J
    T←T+1
    T1←T1+1
    →(T1≤ITER×(ρSITES)[0])/LOOP0
    TEMPLATE←VERTICES
    TEMPLATE2←NVERTICES
    S←(Nρ1)∘.× 2 2 ρ 1 0 0 1
    T←0
    T1←0
    SIG2←(2,N)ρ0.01
    ABOND←(2,N)ρ0.6
   LOOP:T←(ρSITES)[0]|T
    I←SITES[T;0]
    J←SITES[T;1]
    I GLOBAL3 J
    I UPDATE1 J
    T←T+1
    T1←T1+1
    →(T1≤ITER×(ρSITES)[0])/LOOP
    VERTICES←NVERTICES
      ∇
```

```
 ∇ HANDREST3 ITER;T
⍝CARRIES OUT IMAGE RESTORATION OF DEFORMED PICTURE ID
⍝CALLS GLOBAL2   AND UPDATE2
 VERTICES←TEMPLATE←TEMPLATE1
 S←(N⍴1)∘.× 2 2 ⍴ 1 0 0 1
 NVERTICES←TEMPLATE2←TEMPLATE256
 NS←(256⍴1)∘.× 2 2 ⍴ 1 0 0 1
 1 GLOBAL2 0
 1 UPDATE2 0
 TEMPLATE←VERTICES
 TEMPLATE2←NVERTICES
 S←(N⍴1)∘.× 2 2 ⍴ 1 0 0 1
 T←0
 T1←0
LOOP:T←(⍴SITES)[0]|T
 I←SITES[T;0]
 J←SITES[T;1]
 I GLOBAL3 J
 I UPDATE1 J
 T←T+1
 T1←T1+1
 →(T1≤ITER×(⍴SITES)[0])/LOOP
 VERTICES←NVERTICES
 ∇
```

```
∇ HANDREST33 ITER;ITER1;T;T1
⍝CARRIES OUT IMAGE RESTORATION OF DEFORMED PICTURE ID
⍝CALLS GLOBAL2   AND UPDATE2.GLOBAL3 AND UPDATE1
⍝USUS 2- OR 3-STAGE MODEL
 ESTIMATE
 VERT11←(((⍴SITES)[0]+ITER),256,2)⍴0
 VERTICES←TEMPLATE←TEMPLATE1
 S←(N⍴1)∘.× 2 2 ⍴ 1 0 0 1
 NVERTICES←TEMPLATE2←TEMPLATE256
 NS←(256⍴1)∘.× 2 2 ⍴ 1 0 0 1
 1 GLOBAL2 0
 1 UPDATE2 0
 TEMPLATE←VERTICES
 TEMPLATE2←NVERTICES
 VERT11[0;;]←NVERTICES
 S←(N⍴1)∘.× 2 2 ⍴ 1 0 0 1
 T←0
LOOPA:I←SITES[T;0]
 J←SITES[T;1]
 I GLOBAL3 J
 I UPDATE1 J
 VERT11[T+1;;]←NVERTICES
 T←T+1
 →(T<(⍴SITES)[0])/LOOPA
 ITER1←ITER-1
 T←0
 T1←0
LOOP:T←(⍴SITES)[0]|T
 I←SITES[T;0]
 J←SITES[T;1]
 I GLOBAL3 J
 I UPDATE1 J
 T←T+1
 →(T<(⍴SITES)[0])/LOOP
 T1←T1+1
 VERT11[T1+(⍴SITES)[0];;]←NVERTICES
 →(T1<ITER1)/LOOP
∇
```

```
∇ ILEFT GLOBAL5 IRIGHT;PROBS;K;NEWVERTICES;Q;MAXV;MINV;MAXH;MINH
  ;LEFTS;RIGHTS;P;SUBIMAGE;LVERT;LHOR;EDGES;TEMPS
⍝USES GLOBAL SHAPE MODEL WITH GENERATORS FROM ORTHOGONAL GROUP
⍝UPDATES GENERATORS ALONG ARC FROM LEFT HINGE ILEFT TO IRIGHT
⍝NEEDS GLOBAL 2-COL. MATRIX TEMPLATE, N×2×2 ARRAY S,INTEGERS LV A
 ND LH
⍝AND 2×N ARRAYS FOR PARAMETERS SIG2 AND ABOND
⍝PUT GLOBAL VARIABLE RANDOM =1 IF RANDOMIZATION REQUIRED (I.E.
⍝STOCHASTIC RELAXATION) AND =0 FOR MAXIMIZATION
⍝PUT GLOBAL VARIABLE SYNTHESIS=1 FOR PATTERN SYNTHESIS,  AND
⍝=0 FOR IMAGE RESTORATION
⍝IN LAST CASE LV×LH MATRIX ID NEEDED=DEFORMED IMAGE
⍝FOR SYNTHESIS PUT SS=2
⍝BOOLEAN VARIABLE LARGE SHOULD BE SET TO ONE IF LVERT, LHOR ARE L
 ARGE
⍝ALLOWS PART OF PICTURE TO BE MASKED
⍝REMAINING PART OF PICTURE DESCRIBED BY BOOLEAN MATRIX SEEN
 LEFTS←S[N|ILEFT-1;;]
 RIGHTS←S[IRIGHT;;]
 P←1+N|IRIGHT-ILEFT
 CHORD←VERTICES[IRIGHT;]-VERTICES[ILEFT;]
 ACHORD←TEMPLATE[IRIGHT;]-TEMPLATE[ILEFT;]
⍝INCLUDE  IDENTITY ELEMENT IN ARRAY SVALUES
 SVALUES←RANDOMGROUP3 SS-1
 EDGES←TEMPLATE[N|ILEFT+1+⍳P-1;]-TEMPLATE[N|ILEFT+⍳P-1;]
 TEMPS← 0 2 1 ⍉+\[1]+/SVALUES× 0 1 3 2 ⍉((SS-1)⍴1)∘.×EDGES∘.×2⍴1
 SVALUES←SVALUES,[0] S[N|ILEFT+⍳P-1;;]
 TEMPS←(((SS-1)⍴1)∘.×2⍴0),[2] TEMPS
 NEWVERTICES←(((SS-1)⍴1)∘.×VERTICES[ILEFT;]∘.×P⍴1)+TEMPS
 NEWVERTICES←NEWVERTICES,[0]⍉VERTICES[N|ILEFT+⍳P;]
 →SYNTHESIS/SYNTH
 KEEP←∧/(NEWVERTICES[;0;]≥0)∧NEWVERTICES[;0;]<LV-1
 KEEP←KEEP∧∧/(NEWVERTICES[;1;]≥0)∧NEWVERTICES[;1;]<LH-1
 NKEEP←+/KEEP
 NEWVERTICES←NEWVERTICES[KEEP/⍳SS;;]
 SVALUES←SVALUES[KEEP/⍳SS;;;]
 →SYNTHESIS/SYNTH
⍝COUNTER K ENUMERATES SAMPLES
 K←0
 QS←NKEEP⍴0
 MAXV←⌈⌈/⌈/NEWVERTICES[;0;]
 MINV←⌊⌊/⌊/NEWVERTICES[;0;]
 MAXH←⌈⌈/⌈/NEWVERTICES[;1;]
 MINH←⌊⌊/⌊/NEWVERTICES[;1;]
 LVERT←1+MAXV-MINV
 LHOR←1+MAXH-MINH
 VERT←MINV+⍳LVERT
 HOR←MINH+⍳LHOR
 SEENV←(MASKV∊VERT)/MASKV
 SEENH←(MASKH∊HOR)/MASKH
 →(0=(⍴SEENV)×⍴SEENH)/0
 SEEN←(VERT∊SEENV)∘.∧HOR∊SEENH
 SUBIMAGE←ID[VERT;HOR]
LOOP:→(~LARGE)/NORMAL
```

```
⍝BYPASS STATEMENT NORMAL ETC. IF LARGE IS TRUE
 LARGEF
 →CONT
NORMAL:SUMX←DEFPROB⍉NEWVERTICES[K;;]
 DIFF←POS-NEG
 SUMX←+/+/SEEN×DIFF×SUBIMAGE
 SUM1←+/+/SEEN×DIFF
CONT:Q←2×(MIN-MOUT)×SUMX
 QS[K]←Q-((MIN*2)-MOUT*2)×SUM1
 K←K+1
 →(K<NKEEP)/LOOP
⍝TO AVOID OVERFLOW:
 PROBS←*(QS-⌈/QS)÷2×TAU*2
 →RANDOM/RANDOMIZE
 K←PROBS⍳⌈/PROBS
 →SET
RANDOMIZE:PROBS←PROBS÷+/PROBS
 K←+/(1E¯8×?100000000)≥+\PROBS
SET:→SET1
SYNTH:K←0
SET1:S[N|ILEFT+⍳P-1;;]←SVALUES[K;;;]
 VERTICES[N|ILEFT+⍳P;]←⍉NEWVERTICES[K;;]
∇
```

```
     ∇ ILEFT GLOBAL4 IRIGHT;K;Q;MAXV;MINV;MAXH;MINH;LEFTS;RIGHTS;P;SU
        BIMAGE;LVERT;LHOR;EDGES;TEMPS
[1]   ⍝USES GLOBAL SHAPE MODEL WITH GENERATORS FROM ORTHOGONAL GROUP
[2]   ⍝UPDATES GENERATORS ALONG ARC FROM LEFT HINGE ILEFT TO IRIGHT
[3]   ⍝NEEDS GLOBAL 2-COL. MATRIX TEMPLATE. N×2×2 ARRAY S.INTEGERS LV A
        ND LH
[4]   ⍝AND 2×N ARRAYS FOR PARAMETERS SIG2 AND ABOND
[5]   ⍝PUT GLOBAL VARIABLE RANDOM =1 IF RANDOMIZATION REQUIRED (I.E.
[6]   ⍝STOCHASTIC RELAXATION) AND =0 FOR MAXIMIZATION
[7]   ⍝PUT GLOBAL VARIABLE SYNTHESIS=1 FOR PATTERN SYNTHESIS.  AND
[8]   ⍝=0 FOR IMAGE RESTORATION
[9]   ⍝IN LAST CASE LV×LH MATRIX ID NEEDED=DEFORMED IMAGE
[10]  ⍝FOR SYNTHESIS PUT SS=2
[11]  ⍝BOOLEAN VARIABLE LARGE SHOULD BE SET TO ONE IF LVERT. LHOR ARE L
        ARGE
[12]  ⍝EMPLOYS LOCAL ESTIMATION
[13]   LEFTS←S[N|ILEFT-1;;]
[14]   RIGHTS←S[IRIGHT;;]
[15]   P←1+N|IRIGHT-ILEFT
[16]   CHORD←VERTICES[IRIGHT;]-VERTICES[ILEFT;]
[17]   ACHORD←TEMPLATE[IRIGHT;]-TEMPLATE[ILEFT;]
[18]  ⍝INCLUDE  IDENTITY ELEMENT IN ARRAY SVALUES
[19]   SVALUES←RANDOMGROUP3 SS-1
[20]   EDGES←TEMPLATE[N|ILEFT+1+⍳P-1;]-TEMPLATE[N|ILEFT+⍳P-1;]
[21]   TEMPS← 0 2 1 ⍉+\[1]+/SVALUES× 0 1 3 2 ⍉((SS-1)⍴1)∘.×EDGES∘.×2⍴1
[22]   SVALUES←SVALUES,[0] S[N|ILEFT+⍳P-1;;]
[23]   TEMPS←(((SS-1)⍴1)∘.×2⍴0),[2] TEMPS
[24]   NEWVERTICES←(((SS-1)⍴1)∘.×VERTICES[ILEFT;]∘.×P⍴1)+TEMPS
[25]   NEWVERTICES←NEWVERTICES,[0]⍉VERTICES[N|ILEFT+⍳P;]
[26]   →SYNTHESIS/SYNTH
[27]   KEEP←∧/(NEWVERTICES[;0;]≥0)∧NEWVERTICES[;0;]<LV-1
[28]   KEEP←KEEP∧∧/(NEWVERTICES[;1;]≥0)∧NEWVERTICES[;1;]<LH-1
[29]   NKEEP←+/KEEP
[30]   NEWVERTICES←NEWVERTICES[KEEP/⍳SS;;]
[31]   SVALUES←SVALUES[KEEP/⍳SS;;;]
[32]  ⍝COUNTER K ENUMERATES SAMPLES
[33]   K←0
[34]   ACCEPT←QS←NKEEP⍴0
[35]   MAXV←⌈⌈/⌈/NEWVERTICES[;0;]
[36]   MINV←⌊⌊/⌊/NEWVERTICES[;0;]
[37]   MAXH←⌈⌈/⌈/NEWVERTICES[;1;]
[38]   MINH←⌊⌊/⌊/NEWVERTICES[;1;]
[39]   LVERT←1+MAXV-MINV
[40]   LHOR←1+MAXH-MINH
[41]   VERT←MINV+⍳LVERT
[42]   HOR←MINH+⍳LHOR
[43]   VTEMP←VERTICES[N|IRIGHT+⍳N|1+ILEFT-IRIGHT;]
[44]   LOCALEST VTEMP
```

```
LOOP:SUBIMAGE←ID[VERT;HOR]
 SUMX←DEFPROB⍉NEWVERTICES[K;;]
 SUMX2←+/+/(POS-NEG)×ID[VERT;HOR]*2
 SUM1←+/+/POS-NEG
 NIN←SUMT1+SUM1
 NOUT←(LVERT×LHOR)-NIN
 VIN←SUMTX2+SUMX2-((SUMTX+SUMX)*2)÷NIN
 VIN←VIN÷NIN
 VOUT←+/+/ID[VERT;HOR]*2
 VOUT←VOUT-SUMX2+SUMTX2
 VOUT←VOUT-(((+/+/ID[VERT;HOR])-SUMX+SUMTX)*2)÷NOUT
 VOUT←VOUT÷NOUT
⍝TO DEAL WITH INTERSECTION
 VIN←VIN+1000000×VIN≤0
 VOUT←VOUT+1000000×VOUT≤0
 MIN←(SUMTX+SUMX)÷NIN
 MOUT←((+/+/ID[VERT;HOR])-SUMTX+SUMX)÷NOUT
⍝TO AVOID UNDERFLOW
 ACCEPT[K]←MIN<MOUT
 QS[K]←-0.5×(NIN×⍟VIN)+NOUT×⍟VOUT
 K←K+1
 →(K<NKEEP)/LOOP
 PROBS←ACCEPT×*QS-⌈/QS
 →RANDOM/RANDOMIZE
 K←PROBS⍳⌈/PROBS
 →SET
RANDOMIZE:PROBS←PROBS÷+/PROBS
 K←+/(1E¯8×?100000000)≥+\PROBS
SET:→SET1
SYNTH:K←0
SET1:S[N|ILEFT+⍳P-1;;]←SVALUES[K;;;]
 VERTICES[N|ILEFT+⍳P;]←⍉NEWVERTICES[K;;]
∇
```

```
      ∇ ILEFT GLOBAL3 IRIGHT;PROBS;K;NEWVERTICES;Q;MAXV;MINV;MAXH;MINH
         ;LEFTS;RIGHTS;P;SUBIMAGE;LVERT;LHOR;EDGES;TEMPS
[1]    ⍝USES GLOBAL SHAPE MODEL WITH GENERATORS FROM ORTHOGONAL GROUP
[2]    ⍝UPDATES GENERATORS ALONG ARC FROM LEFT HINGE ILEFT TO IRIGHT
[3]    ⍝NEEDS GLOBAL 2-COL. MATRIX TEMPLATE, N×2×2 ARRAY S,INTEGERS LV A
        ND LH
[4]    ⍝AND 2×N ARRAYS FOR PARAMETERS SIG2 AND ABOND
[5]    ⍝PUT GLOBAL VARIABLE RANDOM =1 IF RANDOMIZATION REQUIRED (I.E.
[6]    ⍝STOCHASTIC RELAXATION) AND =0 FOR MAXIMIZATION
[7]    ⍝PUT GLOBAL VARIABLE SYNTHESIS=1 FOR PATTERN SYNTHESIS,  AND
[8]    ⍝=0 FOR IMAGE RESTORATION
[9]    ⍝IN LAST CASE LV×LH MATRIX ID NEEDED=DEFORMED IMAGE
[10]   ⍝FOR SYNTHESIS PUT SS=2
[11]   ⍝BOOLEAN VARIABLE LARGE SHOULD BE SET TO ONE IF LVERT, LHOR ARE L
        ARGE
[12]    LEFTS←S[N|ILEFT-1;;]
[13]    RIGHTS←S[IRIGHT;;]
[14]    P←1+N|IRIGHT-ILEFT
[15]    CHORD←VERTICES[IRIGHT;]-VERTICES[ILEFT;]
[16]    ACHORD←TEMPLATE[IRIGHT;]-TEMPLATE[ILEFT;]
[17]   ⍝INCLUDE  IDENTITY ELEMENT IN ARRAY SVALUES
[18]    SVALUES←RANDOMGROUP3 SS-1
[19]    EDGES←TEMPLATE[N|ILEFT+1+⍳P-1;]-TEMPLATE[N|ILEFT+⍳P-1;]
[20]    TEMPS← 0 2 1 ⍉+\[1]+/SVALUES× 0 1 3 2 ⍉((SS-1)ρ1)∘.×EDGES∘.×2ρ1
[21]    SVALUES←SVALUES,[0] S[N|ILEFT+⍳P-1;;]
[22]    TEMPS←(((SS-1)ρ1)∘.×2ρ0),[2] TEMPS
[23]    NEWVERTICES←(((SS-1)ρ1)∘.×VERTICES[ILEFT;]∘.×Pρ1)+TEMPS
[24]    NEWVERTICES←NEWVERTICES,[0]⍉VERTICES[N|ILEFT+⍳P;]
[25]    →SYNTHESIS/SYNTH
[26]    KEEP←∧/(NEWVERTICES[;0;]≥0)∧NEWVERTICES[;0;]<LV-1
[27]    KEEP←KEEP∧∧/(NEWVERTICES[;1;]≥0)∧NEWVERTICES[;1;]<LH-1
[28]    NKEEP←+/KEEP
[29]    NEWVERTICES←NEWVERTICES[KEEP/⍳SS;;]
[30]    SVALUES←SVALUES[KEEP/⍳SS;;;]
[31]    →SYNTHESIS/SYNTH
[32]   ⍝COUNTER K ENUMERATES SAMPLES
[33]    K←0
[34]    QS←NKEEPρ0
[35]    MAXV←⌈⌈/⌈/NEWVERTICES[;0;]
[36]    MINV←⌊⌊/⌊/NEWVERTICES[;0;]
[37]    MAXH←⌈⌈/⌈/NEWVERTICES[;1;]
[38]    MINH←⌊⌊/⌊/NEWVERTICES[;1;]
[39]    LVERT←1+MAXV-MINV
[40]    LHOR←1+MAXH-MINH
[41]    VERT←MINV+⍳LVERT
[42]    HOR←MINH+⍳LHOR
[43]    SUBIMAGE←ID[VERT;HOR]
```

```
[44] LOOP:→(~LARGE)/NORMAL
[45] ⍝BYPASS STATEMENT NORMAL ETC. IF LARGE IS TRUE
[46]  LARGEF
[47]  →CONT
[48] NORMAL:SUMX←DEFPROB⍉NEWVERTICES[K;;]
[49]  SUBIMAGE←(LVERT,LHOR)⍴1
[50]  SUM1←DEFPROB⍉NEWVERTICES[K;;]
[51] CONT:Q←2×(MIN-MOUT)×SUMX
[52]  QS[K]←Q-((MIN*2)-MOUT*2)×SUM1
[53]  K←K+1
[54]  →(K<NKEEP)/LOOP
[55] ⍝TO AVOID OVERFLOW:
[56]  PROBS←*(QS-⌈/QS)÷2×TAU*2
[57]  →RANDOM/RANDOMIZE
[58]  K←PROBS⍳⌈/PROBS
[59]  →SET
[60] RANDOMIZE:PROBS←PROBS÷+/PROBS
[61]  K←+/(1E¯8×?100000000)≥+\PROBS
[62] SET:→SET1
[63] SYNTH:K←0
[64] SET1:S[N|ILEFT+⍳P-1;;]←SVALUES[K;;;]
[65]  VERTICES[N|ILEFT+⍳P;]←⍉NEWVERTICES[K;;]
     ∇
```

```
     ∇ ILEFT GLOBAL2 IRIGHT;PROBS;K;NEWVERTICES;Q;MAXV;MINV;MAXH;MINH
       ;LEFTS;RIGHTS;P;SUBIMAGE;LVERT;LHOR
[1]   ⍝USES GLOBAL SHAPE MODEL
[2]   ⍝UPDATES GENERATORS ALONG ARC FROM LEFT HINGE ILEFT TO IRIGHT
[3]   ⍝NEEDS GLOBAL 2-COL. MATRIX TEMPLATE, N×2×2 ARRAY S,INTEGERS LV A
       ND LH
[4]   ⍝AND NVECTORS SIGMAS,RHOS
[5]   ⍝S-ARRAY CONTAINS GROUP ELEMENTS
[6]   ⍝FROM GL(2)
[7]   ⍝PUT GLOBAL VARIABLE RANDOM =1 IF RANDOMIZATION REQUIRED (I.E.
[8]   ⍝STOCHASTIC RELAXATION) AND =0 FOR MAXIMIZATION
[9]   ⍝PUT GLOBAL VARIABLE SYNTHESIS=1 FOR PATTERN SYNTHESIS,  AND
[10]  ⍝=0 FOR IMAGE RESTORATION
[11]  ⍝IN LAST CASE LV×LH MATRIX ID NEEDED=DEFORMED IMAGE
[12]  ⍝FOR SYNTHESIS PUT SS=2
[13]  ⍝BOOLEAN VARIABLE LARGE SHOULD BE SET TO ONE IF LVERT, LHOR ARE L
       ARGE
[14]   LEFTS←S[N|ILEFT-1;;]
[15]   RIGHTS←S[IRIGHT;;]
[16]   P←1+N|IRIGHT-ILEFT
[17]   CHORD←VERTICES[IRIGHT;]-VERTICES[ILEFT;]
[18]   ACHORD←TEMPLATE[IRIGHT;]-TEMPLATE[ILEFT;]
[19]   RHOLEFT←RHOS[ILEFT]
[20]   RHORIGHT←RHOS[IRIGHT]
[21]   SIGMA←0.5×SIGMAS[ILEFT]+SIGMAS[IRIGHT]
[22]  ⍝INCLUDE  IDENTITY ELEMENT IN ARRAY SVALUES
[23]   SVALUES←(RANDOMGROUP2 SS-1),[0] 2 2 ⍴ 1 0 0 1
[24]   ARC←⍉TEMPLATE[N|ILEFT+⍳P;]
[25]   NEWVERTICES←((SS⍴1)∘.×VERTICES[ILEFT;]∘.×P⍴1)+SVALUES+.×ARC-ARC[
       ;0]∘.×P⍴1
[26]   →SYNTHESIS/SYNTH
[27]   KEEP←∧/(NEWVERTICES[;0;]≥0)∧NEWVERTICES[;0;]<LV-1
[28]   KEEP←KEEP∧∧/(NEWVERTICES[;1;]≥0)∧NEWVERTICES[;1;]<LH-1
[29]   NKEEP←+/KEEP
[30]   NEWVERTICES←NEWVERTICES[KEEP/⍳SS;;]
[31]   SVALUES←SVALUES[KEEP/⍳SS;;]
[32]   →SYNTHESIS/SYNTH
[33]  ⍝COUNTER K ENUMERATES SAMPLES
```

```
⍝COUNTER K ENUMERATES SAMPLES
 K←0
 QS←NKEEP⍴0
 MAXV←⌈⌈/⌈/NEWVERTICES[;0;]
 MINV←⌊⌊/⌊/NEWVERTICES[;0;]
 MAXH←⌈⌈/⌈/NEWVERTICES[;1;]
 MINH←⌊⌊/⌊/NEWVERTICES[;1;]
 LVERT←1+MAXV-MINV
 LHOR←1+MAXH-MINH
 VERT←MINV+⍳LVERT
 HOR←MINH+⍳LHOR
 SEENV←(MASKV∊VERT)/MASKV
 SEENH←(MASKH∊HOR)/MASKH
 →(0=(⍴SEENV)×⍴SEENH)/0
 SEEN←(VERT∊SEENV)∘.∧HOR∊SEENH
 SUBIMAGE←ID[VERT;HOR]
LOOP:→(~LARGE)/NORMAL
⍝BYPASS STATEMENT NORMAL ETC. IF LARGE IS TRUE
 LARGEF
 →CONT
NORMAL:SUMX←DEFPROB⍉NEWVERTICES[K;;]
 DIFF←POS-NEG
 SUMX←+/+/SEEN×DIFF×SUBIMAGE
 SUM1←+/+/SEEN×DIFF
CONT:Q←2×(MIN-MOUT)×SUMX
 QS[K]←Q-((MIN*2)-MOUT*2)×SUM1
 K←K+1
 →(K<NKEEP)/LOOP
⍝TO AVOID OVERFLOW:
 PROBS←*(QS-⌈/QS)÷2×TAU*2
 →RANDOM/RANDOMIZE
 K←PROBS⍳⌈/PROBS
 →SET
RANDOMIZE:PROBS←PROBS÷+/PROBS
 K←+/(1E¯8×?100000000)≥+\PROBS
SET:→SET1
SYNTH:K←0
SET1:S[N|ILEFT+⍳P-1;;]←((P-1)⍴1)∘.×SVALUES[K;;]
 VERTICES[N|ILEFT+⍳P;]←⍉NEWVERTICES[K;;]
 ∇
```

```
∇ ILEFT GLOBAL2M IRIGHT;PROBS;K;NEWVERTICES;Q;MAXV;MINV;MAXH;MIN
  H;LEFTS;RIGHTS;P;SUBIMAGE;LVERT;LHOR
⍝USES GLOBAL SHAPE MODEL
⍝UPDATES GENERATORS ALONG ARC FROM LEFT HINGE ILEFT TO IRIGHT
⍝NEEDS GLOBAL 2-COL. MATRIX TEMPLATE. N×2×2 ARRAY S.INTEGERS LV A
 ND LH
⍝AND NVECTORS SIGMAS.RHOS
⍝PUT GLOBAL VARIABLE RANDOM =1 IF RANDOMIZATION REQUIRED (I.E.
⍝STOCHASTIC RELAXATION) AND =0 FOR MAXIMIZATION
⍝PUT GLOBAL VARIABLE SYNTHESIS=1 FOR PATTERN SYNTHESIS.  AND
⍝=0 FOR IMAGE RESTORATION
⍝IN LAST CASE LV×LH MATRIX ID NEEDED=DEFORMED IMAGE
⍝FOR SYNTHESIS PUT SS=2
⍝BOOLEAN VARIABLE LARGE SHOULD BE SET TO ONE IF LVERT. LHOR ARE L
 ARGE
⍝S=GROUP GL(2)
⍝              ASSUMES PART OF PICTURE MASKED AND
⍝SEEN ONLY AS DESCRIBED BY G[OBAL BOOLEAN MATRIX SEEN
 LEFTS←S[N|ILEFT-1;;]
 RIGHTS←S[IRIGHT;;]
 P←1+N|IRIGHT-ILEFT
 CHORD←VERTICES[IRIGHT;]-VERTICES[ILEFT;]
 ACHORD←TEMPLATE[IRIGHT;]-TEMPLATE[ILEFT;]
 RHOLEFT←RHOS[ILEFT]
 RHORIGHT←RHOS[IRIGHT]
 SIGMA←0.5×SIGMAS[ILEFT]+SIGMAS[IRIGHT]
⍝INCLUDE  IDENTITY ELEMENT IN ARRAY SVALUES
 SVALUES←(RANDOMGROUP2 SS-1),[0] 2 2 ⍴ 1 0 0 1
 ARC←⍉TEMPLATE[N|ILEFT+⍳P;]
 NEWVERTICES←((SS⍴1)∘.×VERTICES[ILEFT;]∘.×P⍴1)+SVALUES+.×ARC-ARC[
 ;0]∘.×P⍴1
 →SYNTHESIS/SYNTH
 KEEP←∧/(NEWVERTICES[;0;]≥0)∧NEWVERTICES[;0;]<LV-1
 KEEP←KEEP∧∧/(NEWVERTICES[;1;]≥0)∧NEWVERTICES[;1;]<LH-1
 NKEEP←+/KEEP
 NEWVERTICES←NEWVERTICES[KEEP/⍳SS;;]
 SVALUES←SVALUES[KEEP/⍳SS;;]
```

```
⍝COUNTER K ENUMERATES SAMPLES
 K←0
 QS←NKEEP⍴0
 MAXV←⌈⌈/⌈/NEWVERTICES[;0;]
 MINV←⌊⌊/⌊/NEWVERTICES[;0;]
 MAXH←⌈⌈/⌈/NEWVERTICES[;1;]
 MINH←⌊⌊/⌊/NEWVERTICES[;1;]
 LVERT←1+MAXV-MINV
 LHOR←1+MAXH-MINH
 VERT←MINV+⍳LVERT
 HOR←MINH+⍳LHOR
 SUBIMAGE←ID[VERT;HOR]
LOOP:→(~LARGE)/NORMAL
⍝BYPASS STATEMENT NORMAL ETC. IF LARGE IS TRUE
 LARGEF
 →CONT
NORMAL:SUMX←DEFPROB⍉NEWVERTICES[K;;]
 SUBIMAGE←(LVERT,LHOR)⍴1
 SUM1←DEFPROB⍉NEWVERTICES[K;;]
CONT:Q←2×(MIN-MOUT)×SUMX
 QS[K]←Q-((MIN*2)-MOUT*2)×SUM1
 K←K+1
 →(K<NKEEP)/LOOP
⍝TO AVOID OVERFLOW:
 PROBS←*(QS-⌈/QS)÷2×TAU*2
 →RANDOM/RANDOMIZE
 K←PROBS⍳⌈/PROBS
 →SET
RANDOMIZE:PROBS←PROBS÷+/PROBS
 K←+/(1E¯8×?100000000)≥+\PROBS
SET:→SET1
SYNTH:K←0
SET1:S[N|ILEFT+⍳P-1;;]←((P-1)⍴1)∘.×SVALUES[K;;]
 VERTICES[N|ILEFT+⍳P;;]←⍉NEWVERTICES[K;;]
∇
```

```
 ∇PLOTT[⎕]∇
∇ PLOTT
VRTSD←1 VERTEXSD 0
⎕IO←1
20 35 PLOT VRTSD[1;] AND VRTSD[3;] VS⍳35
⎕IO←0
⍳0
'ABOND CONSTANT = ',⍕ABOND[0;0]
'SIG2 CONSTANT = ',⍕SIG2[0;0]
⍳0
⎕PP←3
VRTSD
⍳0
⍳0
⎕PP←7
∇
>
```

```
      ∇ ESTIMATE;SAMPLEIN;SAMPLEOUT;NIN;NOUT
[1]   ⍝ESTIMATES MIN,MOUT, AND TAU FROM DEFORMED IMAGE ID
[2]   ⍝USES INNER MASK CALLED MASKIN, AS BOOLEAN MATRIX,
[3]   ⍝AND OUTER MASK MASKOUT
[4]    SAMPLEIN←(,MASKIN)/,ID
[5]    SAMPLEOUT←(,MASKOUT)/,ID
[6]    NIN←⍴SAMPLEIN
[7]    NOUT←⍴SAMPLEOUT
[8]    MIN←(+/SAMPLEIN)÷NIN
[9]    MOUT←(+/SAMPLEOUT)÷NOUT
[10]   TAU←(+/(SAMPLEIN-MIN)*2)++/(SAMPLEOUT-MOUT)*2
[11]   TAU←(TAU÷NIN+NOUT)*0.5
      ∇

      ∇ ESTIMATEM;SAMPLEIN;SAMPLEOUT;NIN;NOUT
[1]   ⍝ESTIMATES MIN,MOUT, AND TAU FROM DEFORMED IMAGE ID
[2]   ⍝USES INNER MASK CALLED MASKIN, AS BOOLEAN MATRIX,
[3]   ⍝AND OUTER MASK MASKOUT
[4]   ⍝BUT ONLY INSIDE REGION DESCRIBED BY GLOBAL
[5]   ⍝BOOLEAN MATRIX CALLED SEEN
[6]    SEEN←((⍳LV)∊MASKV)∘.∧(⍳LH)∊MASKH
[7]    SAMPLEIN←(,MASKIN∧SEEN)/,ID
[8]    SAMPLEOUT←(,MASKOUT∧SEEN)/,ID
[9]    NIN←⍴SAMPLEIN
[10]   NOUT←⍴SAMPLEOUT
[11]   MIN←(+/SAMPLEIN)÷NIN
[12]   MOUT←(+/SAMPLEOUT)÷NOUT
[13]   TAU←(+/(SAMPLEIN-MIN)*2)++/(SAMPLEOUT-MOUT)*2
[14]   TAU←(TAU÷NIN+NOUT)*0.5
      ∇
```

```
      ∇ L←CHOLESKYR A;N;I;C;J
[1]   ⍝ CHOLESKY DECOMPOSITION OF A SYMMETRIC POSITIVE DEFINITE MATRIX
[2]   ⍝ CALLS LTR1
[3]   ⍝ LAST UPDATE: 6/19/79
[4]    N←(¯1+((1+8×⍴A)*0.5))÷2
[5]    L←A[1]*0.5
[6]    I←1
[7]   S2:→0×⍳N<I←1+J←I
[8]    C←L LTR1(J↑(((I-1)×I)÷2)↓A)
[9]    L←L,C,(A[(I×I+1)÷2]-C+.×C)*0.5
[10]   →S2
      ∇
```

```
      ∇ Z←COARSE
[1]    I←0
[2]    Z←⍳0
[3]   LOOP:Z←Z,(256|NEWLAND[N|I+1]-NEWLAND[I])⍴I
[4]    I←I+1
[5]    →(I<N)/LOOP
      ∇
```

```
      ∇ VA←CONVERT A
[1]   ⍝ CONVERTS A SYMMETRIC MATRIX INTO A VECTOR
[2]   ⍝ LAST UPDATE: 6/19/79
[3]    VA←(,VB∘.≥VB←⍳(1↑⍴A))/,A
      ∇
```

```
      ∇ SUM←DEFPROB VS1;V2;BOOLEAN;COND;TEMP
[1]   ⍝COMPUTES CONTOUR INTEGRALS   NEEDED FOR     DEFORMATION PROBABIL
      ITY GIVEN MATRIX SUBIMAGE
[2]   ⍝OF SIZE LHOR×LVERT AND 3-DIM. ARRAY VERTICES WITH NO  INTEGER VA
      LUES
[3]   ⍝FOR THE COORDINATES (IN ABSOLUTE COORDINATES IN RANGE 0,RESOL)
[4]   ⍝NEEDS GLOBAL VARIABLES HOR AND VERT
[5]   ⍝INDEX ORIGIN 0
[6]    VS2←1⊖[0] VS1
[7]    BOOLEAN←(((VS1[;0]⌈VS2[;0])∘.≥VERT)∧((VS1[;0]⌊VS2[;0])∘.≤VERT))∘
      .∧LHOR⍴1
[8]    TEMP←VS2[;0]-VS1[;0]
[9]    COND←(VS1[;1]∘.×LVERT⍴1)+(VS1[;0]∘.-VERT)×((VS1[;1]-VS2[;1])÷
      TEMP+0=TEMP)∘.×LVERT⍴1
[10]   COND←COND∘.≥HOR
[11]   BOOLEAN←BOOLEAN∧COND
[12]   POS←+/[0]((TEMP<0)∘.∧(LVERT⍴1)∘.∧(LHOR⍴1))∧BOOLEAN
[13]   NEG←+/[0]((TEMP>0)∘.∧(LVERT⍴1)∘.∧(LHOR⍴1))∧BOOLEAN
[14]   SUM←+/+/SUBIMAGE×POS-NEG
      ∇
```

Appendix B
Instructions for Using the Software

Our experiments have used pictures of size $LV = 128$ by $LH = 120$, which is certainly adequate. It would be enough to choose the size 64×60, and even 32×30 would suffice if the patterns are simpler than hands.

The code has been executed on an IBM 3090 under VS APL. The analysis part of the programs could be run on smaller machines, assuming enough memory to be able to hold the 3- or 4-dimensional arrays created during execution.

We assume a template given as a 256×2 matrix TEMPLATE256 and a coarsened version TEMPLATE1 obtained as

$$TEMPLATE1 \leftarrow TEMPLATE256[NEWLAND]$$

where the vector NEWLAND holds the site numbers of the hinges in TEMPLATE256. We have used NEWLAND of lengths 16, 36, and 46; see sections 2 and 4.

To carry out pattern synthesis put $SS \leftarrow 2$, SYNTHESIS=RANDOM← 1 and LARGE← 0. The GLOBAL functions update over a sweep area from site ILEFT to IRIGHT. A convenient way of defining sweep strategies is to define the 2-column matrix SITES and use for example HANDSYNTH3. The reader is encouraged to experiment with different organization of analysis programs and also to vary both the variance like vectors and 2-row matrices respectively, SIGMAS and SIG2 and with the vectors and 2-row matrices RHOS and ABOND. Some values for them have been mentioned in the text but a good deal of experimentation is needed in order to get some intuitive understanding of the role of these parameters in the prior measure.

To carry out pattern analysis, here image reconstruction, requires decisions about the noise model.

The simplest case is to assume, no doubt unrealistically, that the visual noise outside the hand is independent $N(m_{out}, \tau^2)$ and inside the hand independent $N(m_{in}, \tau^2)$. One then uses GLOBAL3 .

If the noise structure is believed to vary more over the picture, which is often the case, local estimation is motivated which is done by GLOBAL4.

If part of the picture is not observable, but hidden by a mask, one should call GLOBAL5.

The program GLOBAL2 assume a rigid transformation from GL(2) between sites ILEFT and IRIGHT. In other words the similarity group element is constant along this arc of the boundary.

GLOBAL2M does the same as GLOBAL2 except that one subset of the picture is assumed to be masked out.

All of these programs operate on N-gons, but the similarity group values, stored in the array S, can be used for interpolating the remaining 256-N vertices. This is done automatically by the programs UPDATE1, UPDATE2.

The result of the GLOBAL-functions, both for synthesis and analysis, consist of the vertices of the N-gon, stored in the global matrix VERTICES, and the similarity group values, stored in the global array S.

The values of S tell us how the arcs in TEMPLATE have been transformed. If $\|s\|$ indicates a norm we can therefore locate abnormalities, suspected or real, by considering the $\|s\|$; $i = 0, 1, \ldots N - 1$.

If global estimation is used the estimated means and standard deviation of the noise is given by the global variables MIN, MOUT, TAU.

Occasionally an updated boundary of the N-gon will fall outside of the allowed area. This will not cause the execution to halt; instead the trial is rejected. This is the way it should be but if the parameters in the prior are set to values corresponding to extremely variable shape, one may wish to be told about it. This can be done by printing the global variable NKEP, telling us how many of the SS trials that were inside the allowed rectangle.

When setting the parameters some guidance can be gotten from the tables in Appendix C.

We have tried to make the code readable rather than optimized. Sometimes, however, we have used concise APL expressions that are less easy to decipher. The reader is then advised to go back to the analytical derivations in the main text, they usually parallel the code closely.

The index origin $\square I0$ is always assumed to have been set to 0.

Appendix C
Variability of Random Shapes

To find out how much variability we can expect for particular settings of the parameters, one can execute the function VERTEXSD. It assumes SIG2 and ABOND given, and has its left and right arguments equal to the left and right side number of the sweep area considered.

The result is a 3-row matrix with x-standard deviations in the first row, the xy-correlation in the second row, and the y-standard deviation in the third row.

Here we show only the (common) standard deviation curves for some values of the parameters with x- and y-components given the same parameter values. Other curves are given in the main text.

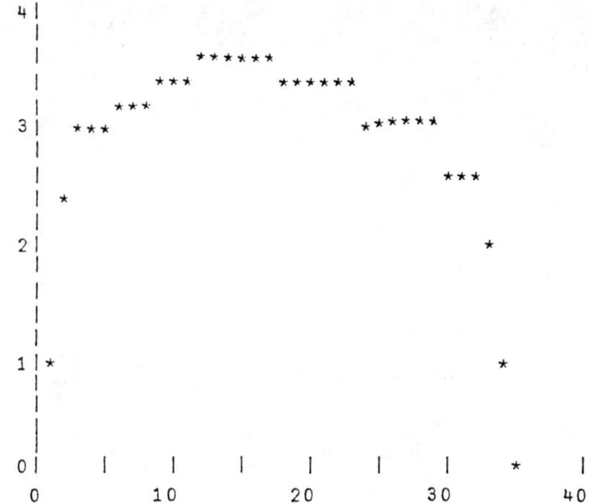

ABOND CONSTANT = 0
SIG2 CONSTANT = 0.005

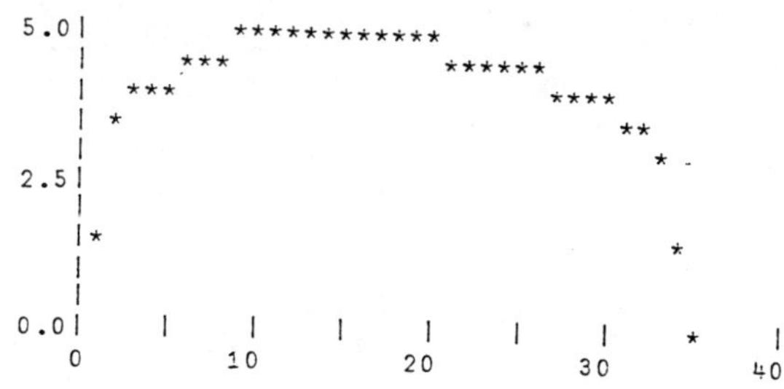

ABOND CONSTANT = 0
SIG2 CONSTANT = 0.01

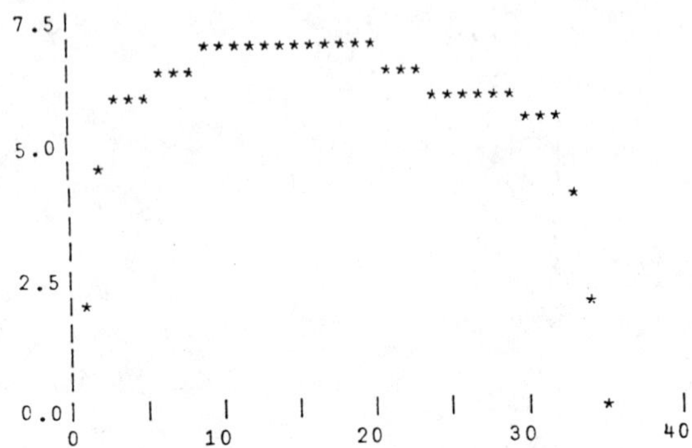

ABOND CONSTANT = 0
SIG2 CONSTANT = 0.02

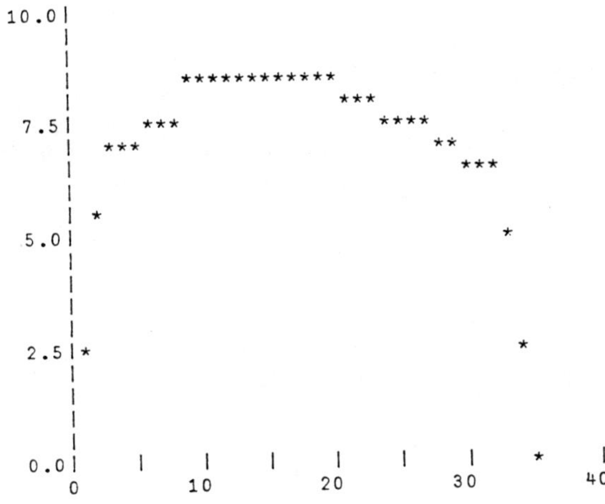

ABOND CONSTANT = 0
SIG2 CONSTANT = 0.03

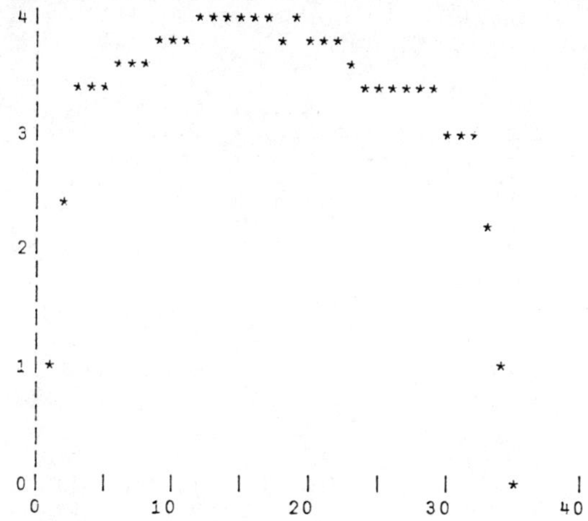

ABOND CONSTANT = 0.2
SIG2 CONSTANT = 0.005

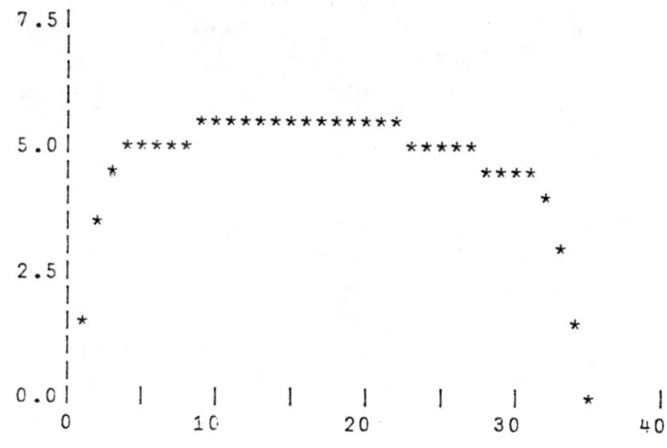

ABOND CONSTANT = 0.2
SIG2 CONSTANT = 0.01

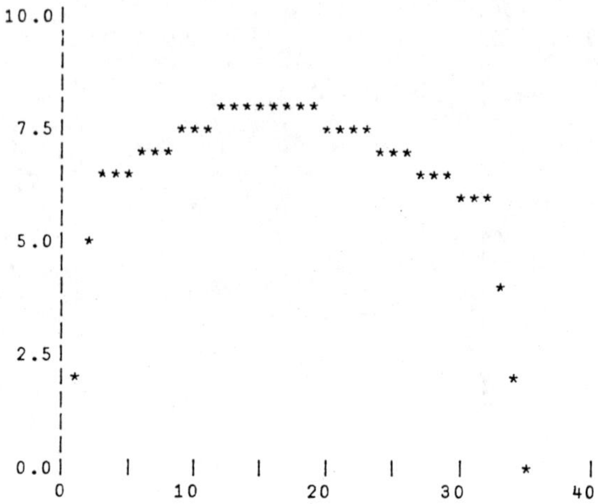

ABOND CONSTANT = 0.2
SIG2 CONSTANT = 0.02

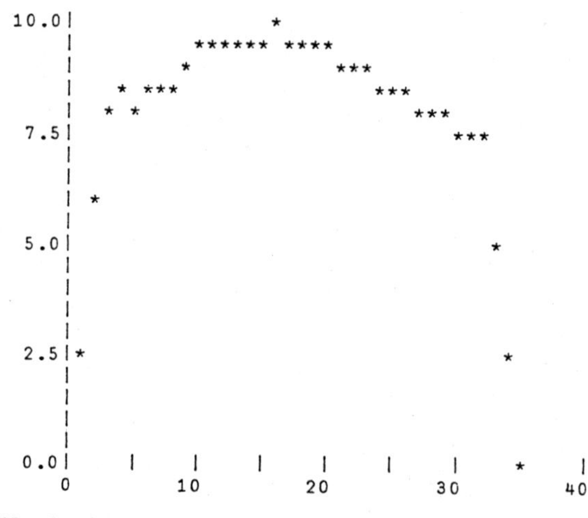

ABOND CONSTANT = 0.2
SIG2 CONSTANT = 0.03

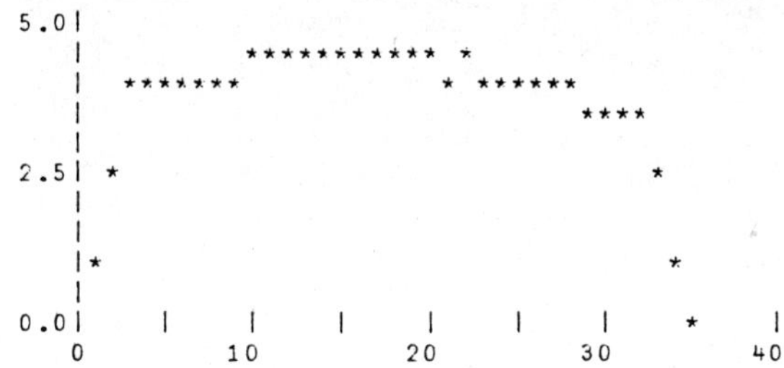

ABOND CONSTANT = 0.4
SIG2 CONSTANT = 0.005

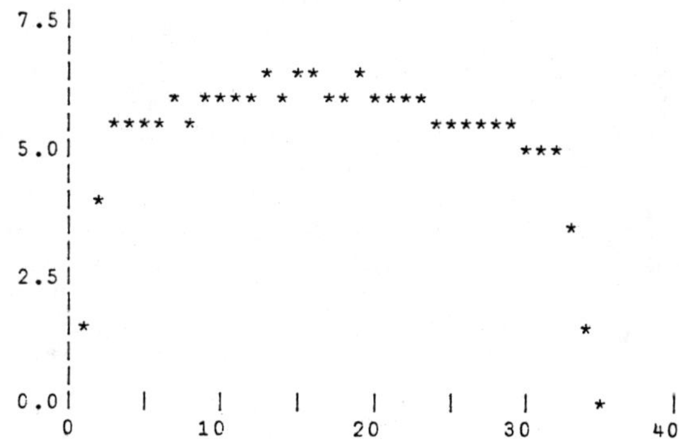

ABOND CONSTANT = 0.4
SIG2 CONSTANT = 0.01

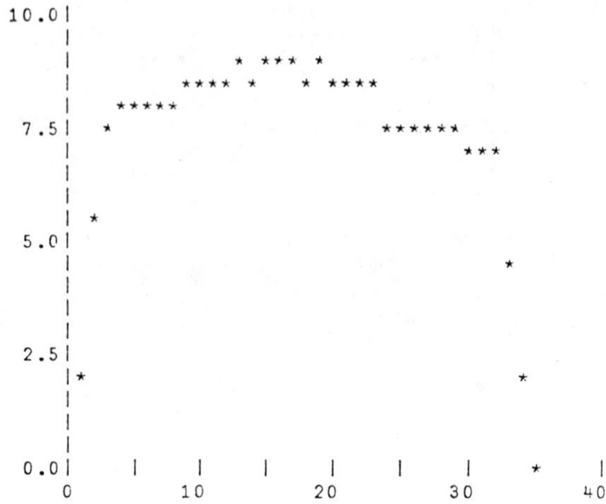

ABOND CONSTANT = 0.4
SIG2 CONSTANT = 0.02

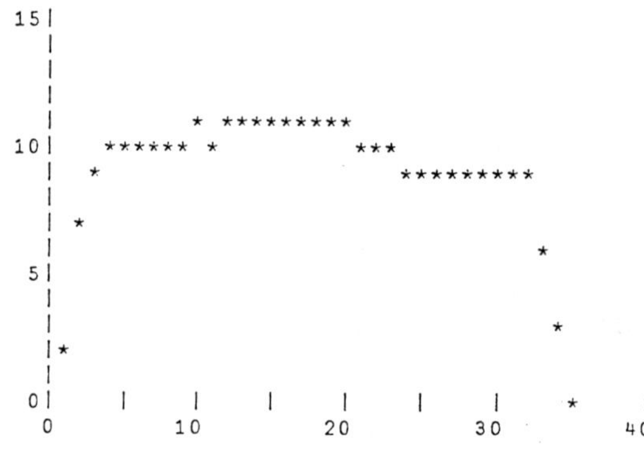

ABOND CONSTANT = 0.4
SIG2 CONSTANT = 0.03

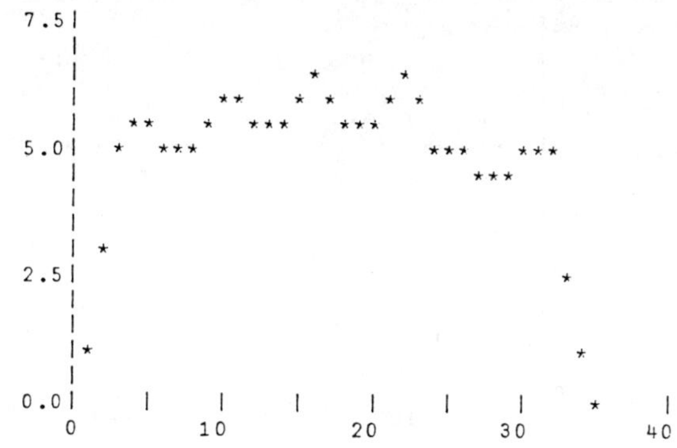

ABOND CONSTANT = 0.8
SIG2 CONSTANT = 0.005

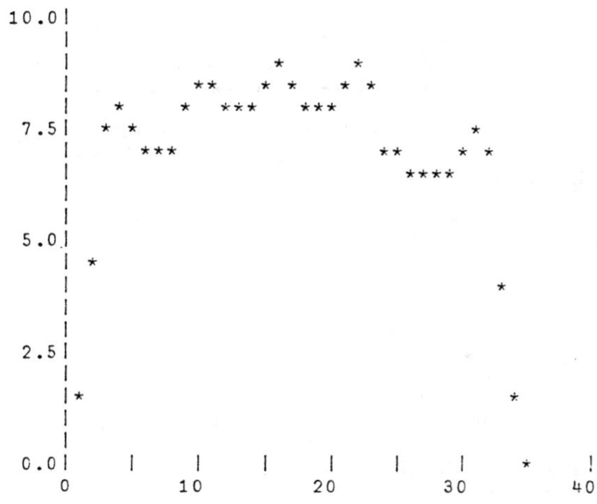

ABOND CONSTANT = 0.8
SIG2 CONSTANT = 0.01

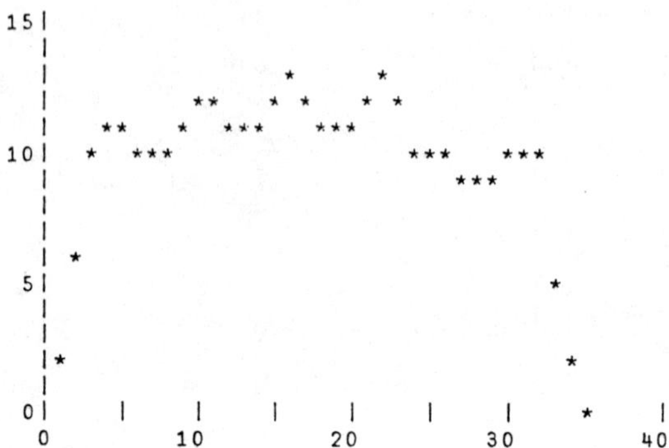

ABOND CONSTANT = 0.8
SIG2 CONSTANT = 0.02

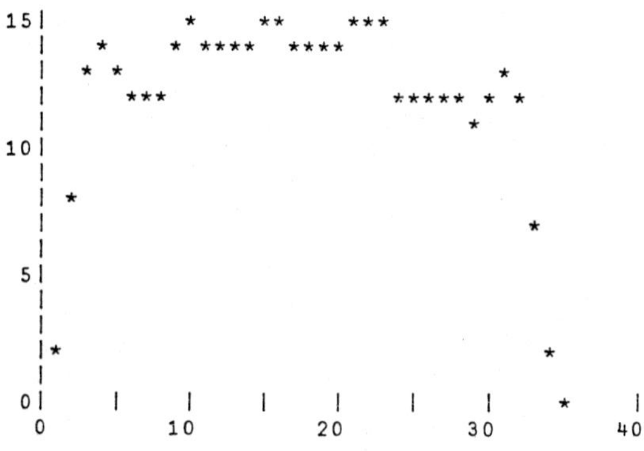

ABOND CONSTANT = 0.8
SIG2 CONSTANT = 0.03

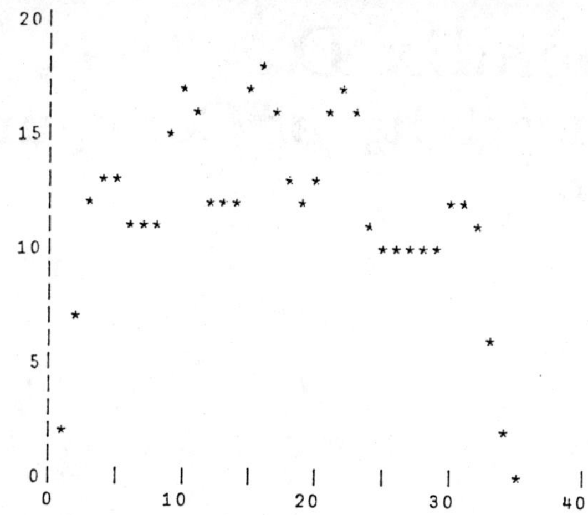

ABOND CONSTANT = 0.95
SIG2 CONSTANT = 0.02

Appendix D
Connectivity of Configuration Space

In this appendix the connectivity of the space of configurations when the group S is uniform scale change, $S = US(2)$, is proven. The proof was not included in Section 2.2 because of its length and complexity.

All polygons discussed in this appendix are assumed to be non-selfintersecting. For each n-gon, C_n, denote clockwise its vertices (all different), angles and edges by $V_i(C_n), \theta_i(C_n)$ and $E_i(C_n), 1 \leq i \leq n$, respectively. The suffix C_n is omitted for the sake of convenience when there is no confusion. We assume V_i is specified once C_n is given. The orientation is taken so that its enclosed domain $\overset{\circ}{C}_n$ lies on the right-hand side of each edge. Let $\mathcal{P}_n$ be the set of all n-gons in the plane with angles in $(0, \pi) \cup (\pi, 2\pi)$. Later, angles of an n-gon are allowed to be π, and we then have $\tilde{\mathcal{P}}_n$, the set of all generalized n-gons. We consider the problem whether two elements C_n, C'_n in $\mathcal{P}_n$ are *equivalent*, in notation $C_n \sim C'_n$, in the sense that they can be transformed within $\mathcal{P}_n$ from one to the other by moving parallel one edge at a time, i.e., there exist an integer m and a function f: $[0,1] \rightarrow \mathcal{P}_n$ such that $f(0) = C_n, f(1) = C'_n$ and for each $1 \leq i < m$, all $f(t), i/m \leq t \leq (i+1)/m$, are obtained from $f(i/m)$ by moving parallel one of its edges. It is clear that $\theta_i(C_n) = \theta_i(C'_n), i \leq i \leq n$, is a necessary condition for C_n, C'_n to be equivalent. The converse, which turns out to be true, is stated as follows:

Theorem 1. *Let* $\mathcal{P}_n(\phi_1, \phi_2, \ldots, \phi_n) = \{C_n \in \mathcal{P}_n : \theta_i(C_n) = \phi_i$ *for* $1 \leq i \leq n\}$, *where* $\phi_i \in 0(,\pi) \cup (\pi, 2\pi)$ *and* $\sum_1^n \phi_i = (n-2)\pi$. *Then (i)* $\mathcal{P}_n(\phi_1, \phi_2, \ldots, \phi_n) \neq \phi$, *and (ii) any two n-gons in it are equivalent.*

Remark: It can be shown that $\mathcal{P}_n(\phi_1, \ldots, \phi_n)$ is representable as a union of convex polyhedral cones in the intersection of an $(n-2)$ dimensional subspace with the positive orthant $\boldsymbol{R}^n_+$. To show connectivity of the resulting (typically non-convex) polyhedral cone one must show not only that appropriate faces of the different cones are glued together, but that they are not joined at a dimension less than $(n-3)$; this approach seemed to be intractable. However this formulation would be appropriate for answering the question of connectivity when $A(\cdot,\cdot)$ has compact support (e.g. Figure 1, Chapter 2), for one could then use the results of Schwartz and Sharir (1983); they develop an abstract algorithm for answering the connectivity question for a space given as a real algebraic variety or more generally where the defining equalities are replaced by inequalities. Using Collins Decomposition they represent the space in a cell decomposition to which an algorithm for graph connectivity could ultimately be applied. This approach could be applied to the case of $A(\cdot,\cdot) > 0$ for all positive scale factors; however, first, such algorithms for all of the various steps do not currently exist, but more importantly, it would only apply to the situation of $\mathcal{P}_n(\phi_1,\ldots,\phi_n)$ for a fixed, specified choice of $(\phi_1,\ldots,\phi_n)$. To prove it simultaneously for all appropriate choices of $(\phi_1,\ldots,\phi_n)$ and n or to where angles as well as lengths are variable, one needs a different approach. The remainder of this section is devoted to such proofs.

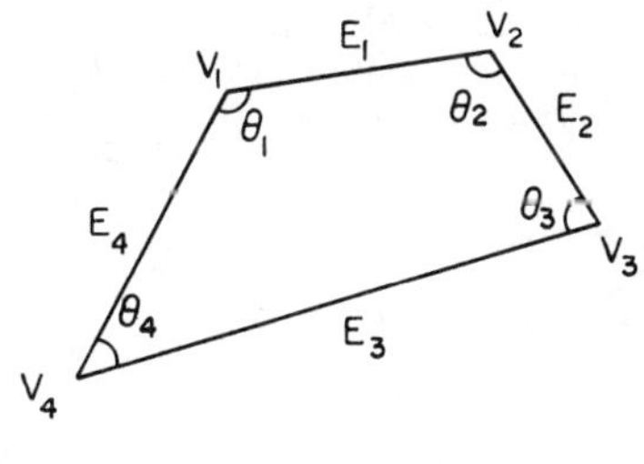

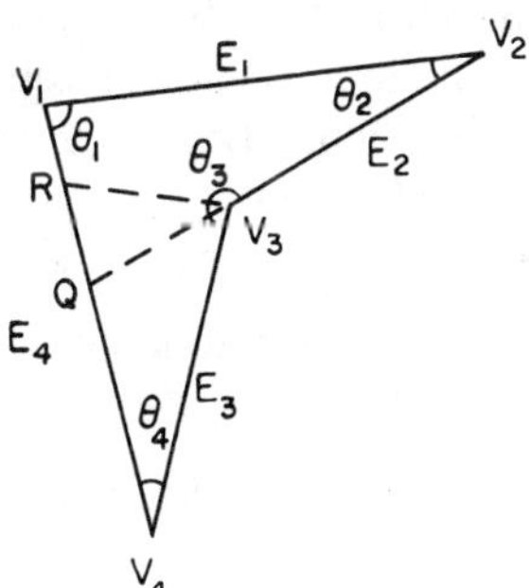

(a) A convex 4-gon (b) A 4-gon with θ a type 1 angle
FIGURE 12

By relaxing the transformation mechanism f, two other weaker equivalent relations are introduced among n-gons. Two n-gons C_n and C'_n in $\mathcal{P}_n$ are called *s-equivalent*, in notation $C_n \overset{s}{\sim} C'_n$, if C_n preserves its local convexity throughout the transform. That is to say, for each $1 \le i \le m, \theta_i(f(t))$ are of the same type for all $0 \le t \le 1$. By definition an angle θ is of type 0 if $0 < \theta < \pi$, and of type 1 if $\pi < \theta < 2\pi$. If the range of f is $\widetilde{\mathcal{P}}_n$ instead of $\mathcal{P}_n$ and angles are allowed to change type so that $\theta_i(f(t)) = \pi$ could hold for certain t, then C_n and C'_n are called *w-equivalent*, in notation $C_n \overset{w}{\sim} C'_n$. Also, notationally, we will denote a k-gon by Δ_k; a 3-gon will be simply denoted by Δ and a 4-gon by $\square$. For example, we have in Figure 12(b) $\square V_1V_2V_3Q \in \widetilde{\mathcal{P}}_4 \setminus \mathcal{P}_4$ and $\square V_1V_2V_3V_4 \overset{w}{\sim} \square V_1V_2V_3Q \overset{w}{\sim} \square V_1V_2V_3R$ by moving edge E_3 along the half line $\overrightarrow{V_4V_1}$ and toward V_1. We will denote a half line by $\rightarrow$ and a line by $\leftrightarrow$. The following results are analogs to Theorem 1.

Theorem 2. *Let $A = (a_1, a_2, \ldots, a_n)$ be a $0-1$ n-tuple. Then any two n-gons in $\mathcal{P}_n(A) = \{C_n \in \mathcal{P}_n : \theta_i(C_n)$ is of type a_i for $1 \leq i \leq n\}$ are s-equivalent.*

Theorem 3. *Any two n-gons in $\widetilde{\mathcal{P}}$ are w-equivalent.*

The proof of Theorem 1 is based on the principle of "divide and conquer". It is clear that $\sum_1^n \phi_i = (n-2)\pi$ is a necessary condition for $\mathcal{P}_n(\phi_1, \phi_2, \ldots, \phi_n) \neq \phi$. That it is also a sufficient condition is proved by using mathematical induction, because one such n-gon can be constructed from adding or cutting a tiny tip (a 3-gon or 4-gon) off some vertex, say the ith one, of a related $(n-1)$-gon C_{n-1}. If Theorem 1 were true, any $C_n \in \mathcal{P}_n(\phi_1, \phi_2, \ldots, \phi_n)$ would be equivalent to C_{n-1} plus or minus a tiny tip located at $V_i(C_{n-1})$. This can be done by properly moving certain parts of C_n. Since a tiny tip can be dealt with as if it does not exist, Theorem 1(ii) follows again from mathematical induction. Then Theorems 2 and 3 can be verified easily by using the results just established.

Several lemmas are needed for the proof of Theorem 1. We first introduce some notations. Let $C_n \in \mathcal{P}_n$ and $\Delta_k \in \mathcal{P}_k$. The sum $C_n \oplus \Delta_k(i)$ of C_n with Δ_k at vertex $V_i(C_n)$ is defined to be the polygon resulted from adding Δ_k to C_n by first translating Δ_k so that $V_1(\Delta_k)$ coincides with $V_i(C_n)$. In the topology terminology, $C_n \oplus \Delta_k(i) = \partial(\mathring{C}_n \cup \mathring{\Delta}_k)$. It is also required that

(1) $$C_n \oplus \Delta_k(i) \text{ is non self-intersecting,}$$

(2) $$\mathring{C}_n \cap \mathring{\Delta}_k = \phi.$$

Hence, Δ_k and C_n are "glued" together along edges $E_{i-1}(C_n)$ and/or $E_i(C_n)$. Similarly, the difference $C_n \ominus \Delta_k(i)$ is defined to be $\partial(\mathring{C}_n \setminus \mathring{\Delta}_k)$ with the requirements that

(3) $$C_n \ominus \Delta_k(i) \text{ is non self-intersecting,}$$

(4) $$\mathring{\Delta}_k \subset \mathring{C}_n.$$

Since in what follows we are interested only in the case that after translation, Δ_k is contained in a small neighborhood of $V_i(C_n)$, it is easy to see that one of the following holds for the sum (difference) case:

(5)
$$\theta_2(\Delta_k)\ (\theta_k(\Delta_k)) < \pi \text{ and } E_1(\Delta_k)\ (E_k(\Delta_k)) \leq E_{i-1}(C_n) \setminus \{V_{i-1}(C_n)\}$$

(6)
$$\theta_k(\Delta_k)\ (\theta_2(\Delta_k)) < \pi \text{ and } E_k(\Delta_k)\ (E_1(\Delta_k)) \leq E_i(C_n) \setminus \{V_{i+1}(C_n)\}.$$

Furthermore, both (5) and (6) hold iff $\theta_1(\Delta_k) + \theta_i(C_n) = 2\pi(\theta_1(\Delta_k) = \theta_i(C_n))$. See Figure 13, where C_4 is given in Figure 12(a).

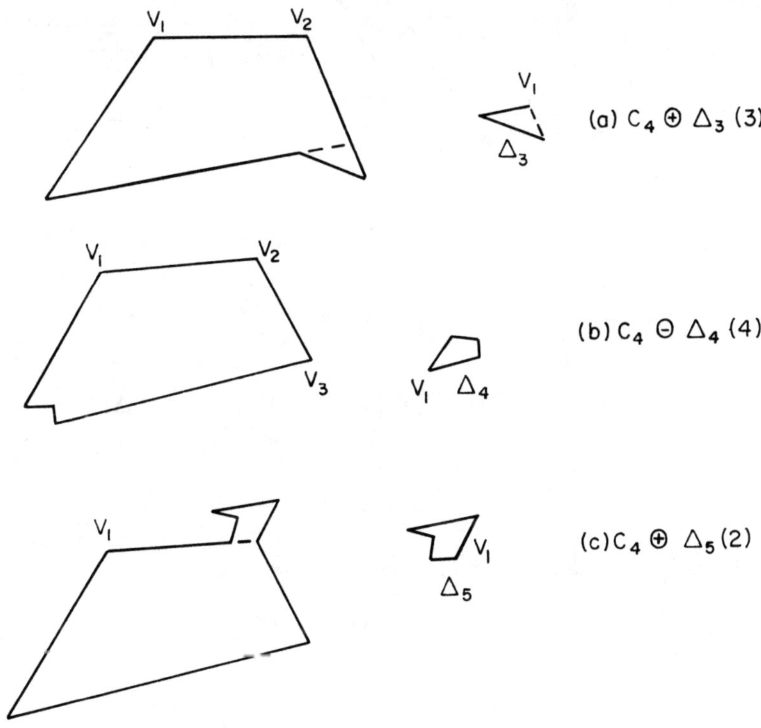

FIGURE 13

The precise meaning of (1) is explained in the following lemma.

Lemma 1. *Let* $\circ = \oplus$ *or* $\ominus$. *Assume (i)* $C_n, C'_n \in \mathcal{P}_n$ *and* $C_n \sim C'_n$, *(ii)* $\Delta_k \in \mathcal{P}_k$ *is sufficiently small and* $C_n \circ \Delta_k(i)$ *is well-defined. Then* $C_n \circ \Delta_k(i) \sim C'_n \circ \Delta_k(i)$.

Proof: For brevity we consider only the case that $\circ = \oplus$ and $i = 2$. the other cases can be treated similarly. By the definition of equivalence, there exist a transformation mechanism f and an integer m such that

$$C_n f(0) \to f(1/m) \to f(2/m) \to \cdots \to f(m/m) = C'_n$$

where for each $0 \le j < m$, only one edge, say $\widetilde{E}(j)$, is moved parallel from $f(j/m)$ to $f((j+1)/m)$.

It suffices to show

$$f(0) \oplus \Delta_k(2) \sim f(1/m) \oplus \Delta_k(2). \tag{7}$$

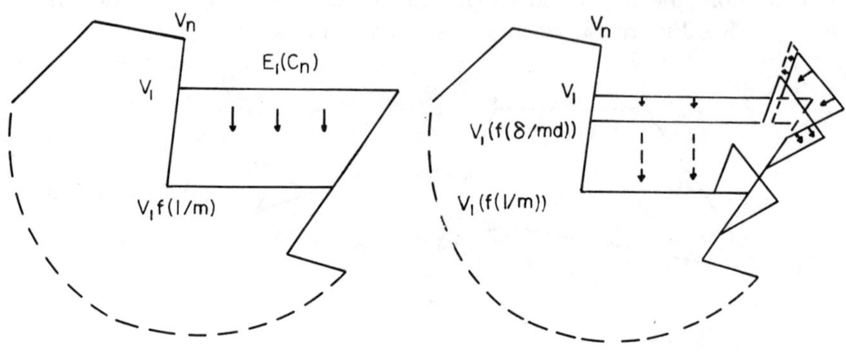

(a) $C_n \to f(1/m)$ (b) $C_n \oplus \Delta_k(2) \sim f(\delta/(md)) \oplus \Delta_k(2)$. Eventually $C_n \oplus \Delta_k(2) \sim f(1/m) \oplus \Delta_k(2)$

FIGURE 14

Because by the same reasoning, we will have

$$C_n \oplus \Delta_k(2) \sim f(1/m) \oplus \Delta_k(2) \sim f(2/m) \oplus \Delta_k(2) \sim \cdots \sim C'_n \oplus \Delta_k(2).$$

Only the case $\widetilde{E}(0) = E_1(C_n)$ or $E_2(C_n)$ needs to be considered. Otherwise, (7) automatically holds as long as the added tip Δ_k is sufficiently small. Taken $\widetilde{E}(0) = E_1(C_n)$ for example. Suppose $E_1(C_n)$ is moved inward a distance d with a constant speed md. See Figure 14(a). It is clear that (7) cannot be reached by moving $E_1(C_n)$ alone. Since a small perturbation is always allowed for a polygon, $f(0) \oplus \Delta_k(2) \sim f(\delta/md) \oplus \Delta_k(2)$ holds for sufficiently small Δ_k and $\delta = d/N$. This can be done by first moving inward $E_1(C_n)$ a distance d and then moving subsequently $E_2(\Delta_k), E_3(\Delta_k), \ldots,$ one by one. See Figure 14(b). If the Δ_k's in both $f(0) \oplus \Delta_k(2)$ and $f(\delta/(md)) \oplus \Delta_k(2)$ are contained in the δ-neighborhood of $V_2(C_n)$ and

$$2\delta < \min_{0 \leq t \leq 1} \; dist. \left(V_2(f(t)), f(t) \setminus \left(\bigcup_{i=1}^{2} E_i(f(t)) \right) \right)$$

the previous procedure can be repeated over and over. Thus, we have in N steps

$$f(0) \oplus \Delta_k(2) \sim f(\delta/(md)) \oplus \Delta_k(2) \sim f(2\delta/(md)) \oplus \Delta_k(2) \\ \sim \cdots \sim f(1/m) \oplus \Delta_k(2).$$

This verifies (7) and the lemma is proved.

Q.E.D.

Remark. It is clear from the previous proof that if the transform from C_n to C'_n is done within a certain domain G, then by first choosing a sufficiently small Δ_k, one has a transform from $C_n \circ \Delta_k(i)$ to $C'_n \circ \Delta_k(i)$.

The following lemmas demonstrate how certain parts of C_n can be moved properly. It is clear that a neighborhood $N(C_n; \delta)$ (in the topology sense) of C_n ($C_n \cup \overset{\circ}{C}_n$ to be precise) is formed by moving all its edges outward a small distance δ. In fact, $\partial N(C_n; \delta) \sim C_n$ if δ is small enough and all edges are moved parallel. A partial neighborhood $N(C_n; \delta; E_i)$ of C_n is obtained as above except that edge $E_i = E_i(C_n)$ is not moved. Similarly, we can define $N(C_n; \delta; E_i, E_{i+1})$ for two adjacent edges. See Figure 15. Denote by $N(P; \epsilon)$ the ϵ neighborhood of point P in the plane.

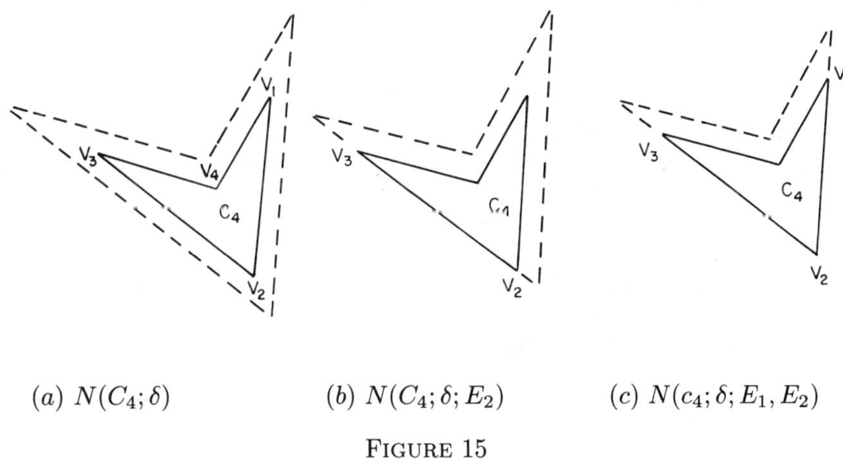

(a) $N(C_4; \delta)$ (b) $N(C_4; \delta; E_2)$ (c) $N(c_4; \delta; E_1, E_2)$

FIGURE 15

Lemma 2. *Let $C_n \in \mathcal{P}_n$ with θ_1, θ_2 being of type 0.*

(i) If θ_3 is of type 0, then for any $N(C_n; \delta; E_1, E_2)$ and any neighborhood $N(V_2; \epsilon), C_n$ can be transformed to an equivalent C'_n such that

(a) $C'_n \subset N(V_2; \epsilon)$,

(b) all the intermediate n-gons have V_2 as their second vertex and are contained in $N(C_n; \delta; E_1, E_2)$. That is to say, the transform is operated completely in $N(C_n; \delta; E_1, E_2)$.

(ii) If θ_3 is of type 1, then the previous results hold with $N(C_n; \delta; E_1, E_2)$ replaced by $N(C_n; \delta; E_1)$ and V_2 replaced by V_1 or not. (See Figures 16(b) and (c).)

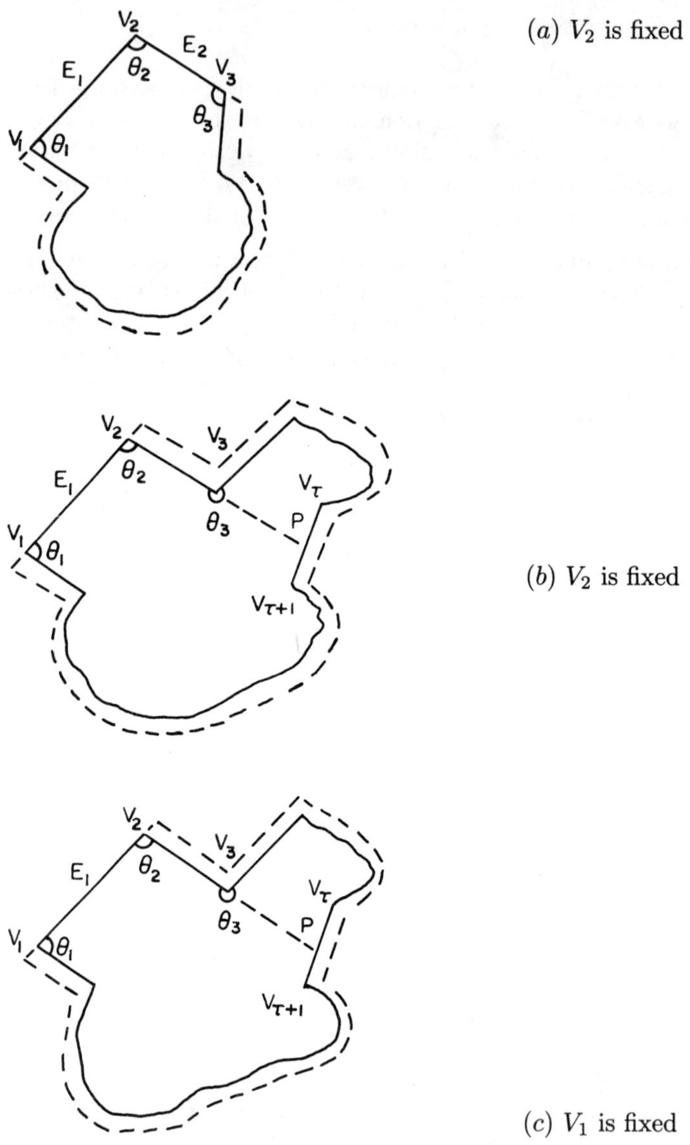

(a) V_2 is fixed

(b) V_2 is fixed

(c) V_1 is fixed

FIGURE 16

Proof. The proof is done by induction on n. Let us first note that C_n is convex iff all its angles are type 0. In that case the lemma follows easily from the mathematical induction and the following observations: (1) The transform can in fact be done inside C_n rather than in $N(C_n; \delta, E_1, E_2)$ required in (b); (2) Replacing edges E_i and E_{i+1} by a new edge $\overline{V_i V_{i+2}}$ we obtain a convex $(n-1)$-gon C_{n-1}. The way edges of C_n are moved is determined by those of C_{n-1}. See Figure 17. Thus the lemma is true for

$n = 3$. Since there are essentially two kinds of 4-gons (see Figure 12), it is easy to see that the lemma holds for $n = 4$.

Suppose the lemma holds for all $k < n$. We now show that it is true for n. The case that C_n is convex has just been treated. In the following we may assume without loss of generality that C_n has at least one angle of type 1.

Case(i). θ_3 is of type 0. Let θ_σ be the first type 1 angle coming after θ_3. Since $\overset{\circ}{C}_n$ is on the right-hand side of $C_n, \partial C_n \cap (\overleftrightarrow{V_{\sigma-1}V_\sigma} \setminus \overrightarrow{V_\sigma V_{\sigma-1}}) \neq \phi$. Let P be the point in this set that is closest to V_σ. By slightly shifting some edge that P is on we may assume $P \in E_\tau$ (for some τ) and is not an endpoint of E_τ. See Figure 18. Similarly we can find a point $Q \in \overset{\circ}{E}_\mu$ such that Q is closest to V_σ in $C_n \cap (\overleftrightarrow{V_\sigma V_{\sigma+1}}) \setminus \overrightarrow{V_\sigma V_{\sigma+1}})$.

We claim that when we travel from V_σ along C_n we reach P before we reach Q. See Figure 19. Otherwise, consider the polygon D formed by $\overline{QV_{\mu+1}} \cup (\bigcup_{\mu<i<\tau} E_i) \cup \overline{V_\tau P} \cup \overline{PV_\sigma} \cup \overline{V_\sigma Q}$. (If $E_\mu = E_\tau$, there is nothing to do in view of the orientation of C_n.) Since $E_{\sigma-1}, E_\sigma$ are not on D, it is clear (see Figure 20) that $V_{\sigma-1}$ and $V_{\tau+1}$ cannot both be in the interior or the exterior of D. As we continue to travel from $V_{\tau+1}$ along C_n we have to cross D before we reach $V_{\sigma-1}$. This contradicts the nonselfintersecting assumption on C_n.

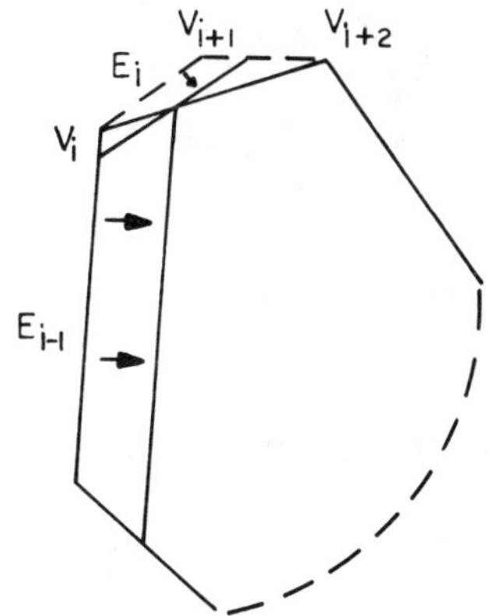

(a) In $C_n k, E_i$ and E_{i-1} are moved

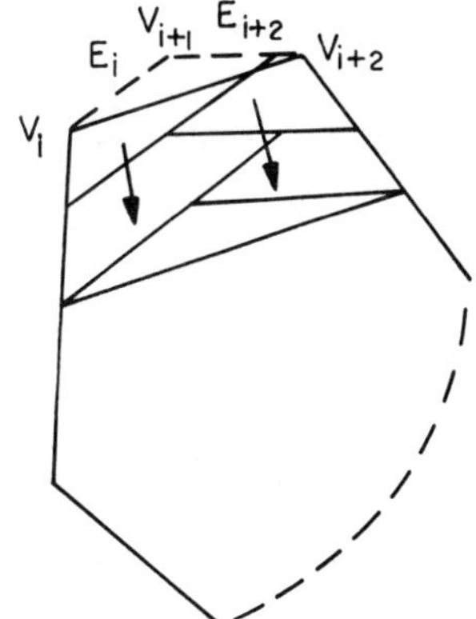

(b) In C_n, the order of move is $E_1, E_{i+1}, E_i, E_{i+1}$

FIGURE 17

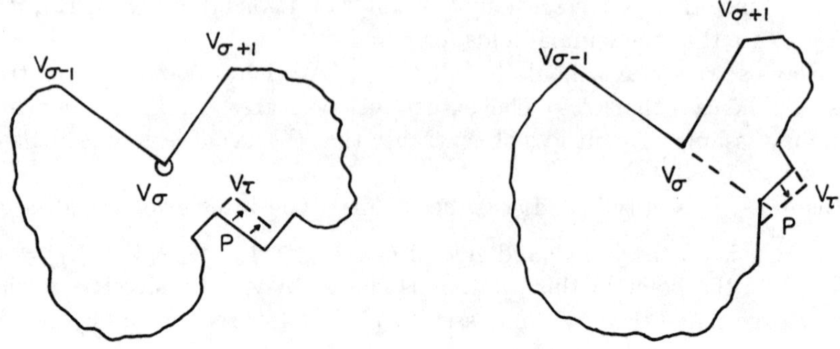

FIGURE 18. P can be made not a vertex of C_n by slightly shifting some edge.

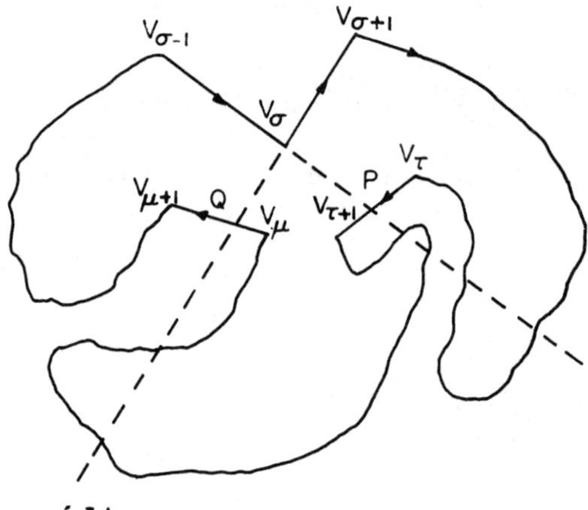

FIGURE 19. $V_\sigma \to P \to Q$ along C_n.

Subcase 1. Both $E_1, E_2 \notin \{E_k : \sigma \leq k \leq \tau\}$. Let Δ be the polygon formed by $(\bigcup_{\sigma\leq k<\tau} E_k) \cup \overline{V_\tau P} \cup \overline{PV_\sigma}$ and $\widetilde{C}$ the polygon by $(\bigcup_{\tau<k<\sigma-1} E_k) \cup \overline{V_{\sigma-1}P} \cup \overline{PV_{\tau+1}}$. Define $P = V_1(\Delta) = V_\sigma(\widetilde{C})$. Then E_1, E_2 are edges of $\widetilde{C}$. It is clear that $\widetilde{C}$, Δ have fewer edges than C_n does and $N(\widetilde{C}; \delta'; E_1, E_2) \subset N(C_n; \delta; E_1, E_2)$ holds for δ' small enough. By using twice the induction hypothesis, $\widetilde{C}$ can be transformed to an equivalent $\widetilde{C}'$ with the properties stated in the lemma, and Δ can be regarded as if $\Delta \subseteq N(P; \epsilon)$, where ϵ can be made arbitrarily small. See Figure 21. Since $C_n \sim \widetilde{C} \oplus \Delta(\sigma)$ the conclusion follows from Lemma 1. See the remark after Lemma 1.

Subcase 2. Both $E_1, E_2 \notin \{E_k : \mu \leq k < (mod\ n) < \sigma\}$. Define $\widetilde{C}$, Δ to

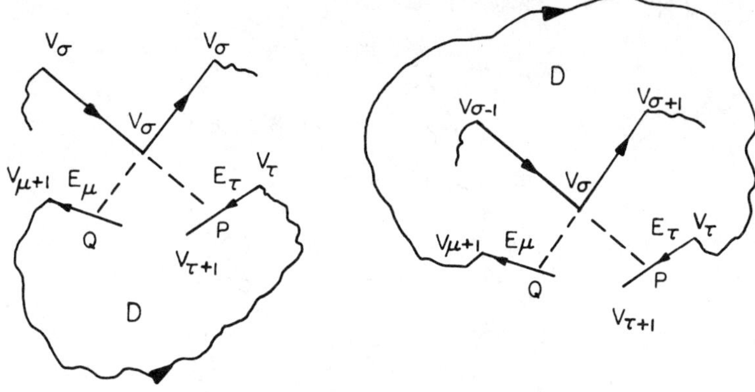

FIGURE 20. $V_\sigma \to Q \to P$ is impossible.

be the polygons formed by $(\bigcup_{\sigma \leq k < \mu} E_k) \cup \overline{V_\mu P} \cup \overline{QV_{\sigma+1}}$ and $(\bigcup_{\mu \leq k < \sigma} E_k) \cup \overline{V_\sigma Q} \cup \overline{QV_{\mu+1}}$ respectively. Then repeat the same argument in Subcase 1.

Subcase 3. $\tau = \mu = 1$. Repeat the same argument in Subcase 1. The only difference is that Lemma 2.5(i) can be applied to Δ. This is due to the fact that $\theta_1 =< V_n V_1 P, < V_1 P V_\sigma$ and $< P V_\sigma V_{\sigma+1}$ are all of type 0. See Figure 22.

Subcase 4. $\tau = \mu = 2$. Repeat the same argument in Subcase 2. See Figure 23.

Subcase 5. $\tau = 1$ and $\mu = 2$. Same as in Subcase 3 or 4. See Figure 24.

Case (ii). θ_3 is of type 1. It is clear from Figure 16(b) that $E_1 \notin \{E_k : 3 \leq k \leq \tau\}$. Now apply the same argument used in Subcase 1 of Case (i).

The lemma is thus proved.

Q.E.D.

Lemma 2 means that we can "cut" C_n along the edges of any type-1 angle. For example, the polygon formed by $(\bigcup_{\sigma \leq k < \tau} E_k) \cup \overline{V_\tau P} \cup \overline{PV_\sigma}$ in Figure 19 can be shrunk along $\overline{V_\sigma P}$ to a tiny tip Δ in $N(P;\epsilon)$. See Figure 21. By Lemma 1 Δ can be regarded as if nonexisting. This reduces the number of edges in C_n and thus paves the way for the mathematical induction. This is the key procedure in the present approach and will be applied many times hereafter.

Lemma 3. *Let $C_n \in P_n$ with θ_2 an angle of type 0 and S be a point outside the sectorial region formed by $< V_1 V_2 V_3$. Then there exists $C_n' \in P_n$ equivalent to C_n with the following properties: (a) The transform is done in a prescribed domain $N(C_n;\delta;E_1,E_2)$; (b) V_2 is fixed throughout the* *transform;* (c) $V_2S \cap C'_n = \phi$.

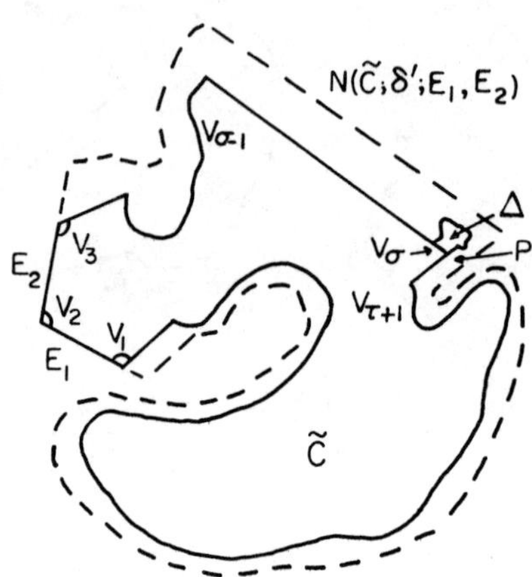

FIGURE 21. Δ can be made arbitrarily small.

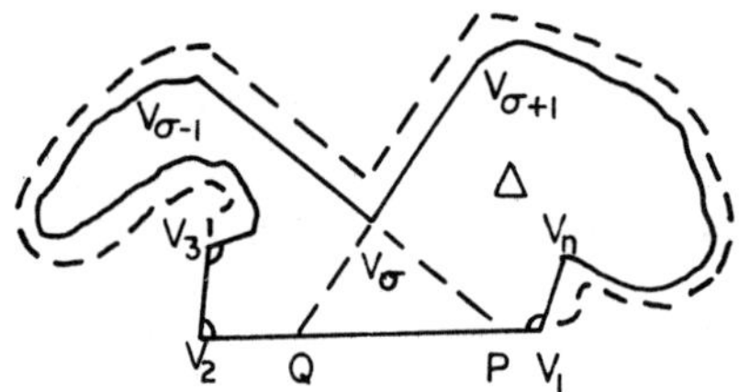

FIGURE 22. Both P, Q are in E_1.

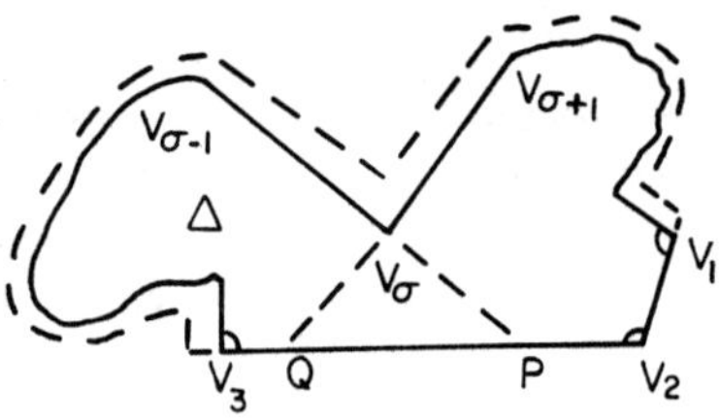

Figure 23. Both P, Q are in E_2.

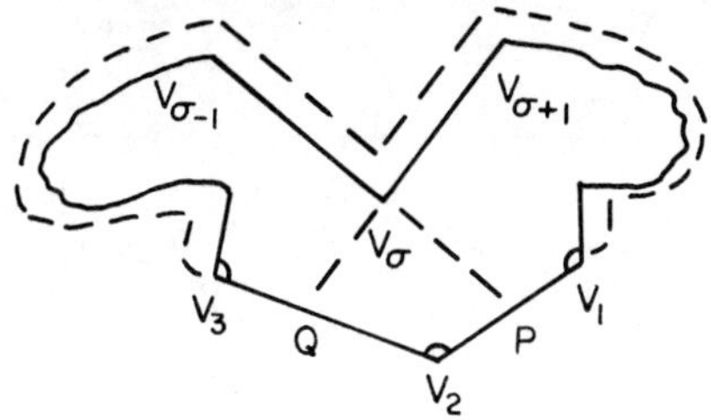

FIGURE 24. Both P, Q are in E_3.

Proof: The proof is done by induction on $n \geq 3$.

Let (x_i, y_i) be the coordinate of V_i. For the convenience of discussion, we assume $\overrightarrow{V}_2S$ is the positive x-axis and $y_3 \leq 0$. See Figure 25(a). The case $y_3 > 0$ can be treated similarly.

If $\overrightarrow{V}_2S \cap C_n = \phi$ then there is nothing to prove. In particular, it is so if C_n is convex. Therefore, we may assume $\overrightarrow{V_2S} \cap C_n \neq \phi$. This implies C_n has at least an angle of type 1. Let E_σ be the first edge crossing $\overrightarrow{V_2S}$ as we travel from V_2 along C_n. By using a small perturbation (see Figure 18) we may assume $P = E_\sigma \cap \overrightarrow{V_2S}$ is not an endpoint of E_σ.

Case 1. Let $y_\tau = \min_{3 \leq k \leq \sigma} y_k$ and have the smallest such subscript. See Figure 25(a). Then θ_τ is of type 1. Define $Q \in E_\mu$ be the point in $C_n \cap (\overleftrightarrow{V_\tau V_{\tau-1}} \backslash \overrightarrow{V_\tau V_{\tau-1}})$ that is most close to V_τ. See Figure 25(a). By Lemmas 1 and 2 the polygon D formed by $\left(\bigcup_{\tau \leq k < \mu} E_k \right) \cup \overline{V_\mu Q} \cup \overline{QV_\tau}$ can be first shrunk and then "neglected" with the properties stated in the Lemma. This reduces C_n to a polygon with fewer edges. The conclusion then follows the induction.

Case 2. $y_\sigma > 0$. Since C_n is nonself-intersecting we have to cross $\overline{V_2P}$ in order to return to V_1 from $V_{\sigma+1}$ along C_n. Assume $T \in E_\ell$ is the last crossing point. See Figure 25(b). Let $y_\tau = \max_{\ell \leq k \leq n+1} y_k$ and have the largest such subscript $(y_{n+1} = y_1)$. Then θ_τ is of type 1. Define $Q \in \overset{\circ}{E}_\mu$ be the point in $C_n \cap (\overleftrightarrow{V_\tau V_{\tau-1}} \backslash \overrightarrow{V_\tau V_{\tau-1}})$ that is most close to V_τ, and D the polygon formed by $\left(\bigcup_{\mu < k < \tau} E_k \right) \cup \overline{V_\tau Q} \cup \overline{QV_{\mu+1}}$. Now proceed as in Case 1.

Q.E.D.

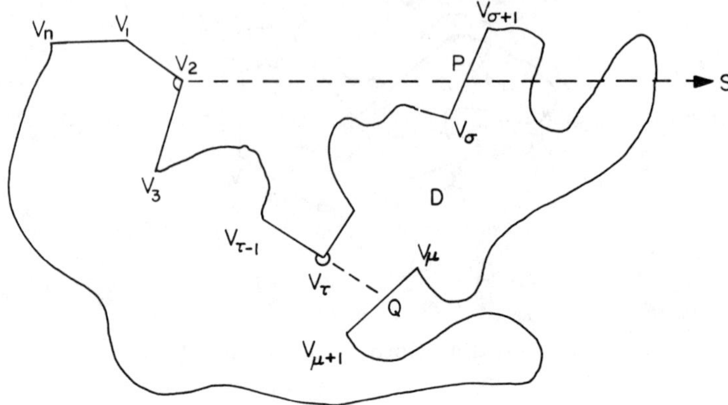

(a) Crossing from below first.

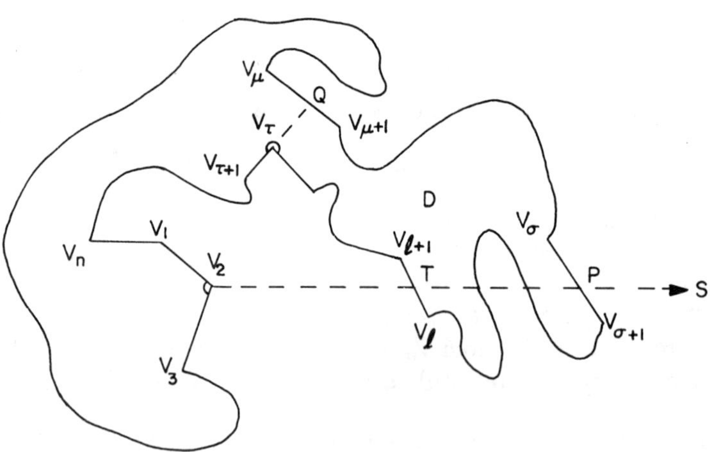

(b) Crossing from above first.

FIGURE 25. D can be shrunk arbitrarily close to Q.

Remark. The proof of Lemma 3 tells us more about the transform from C_n to C'_n. For example, if θ_1 is of type 0 and $\overleftrightarrow{V_nV_1} \parallel \overleftrightarrow{V_2S}$ (i.e., are parallel), then (b) in Lemma 3 can be replaced by: (b$'$) throughout the transform V_n, V_1, V_2 are fixed, and $E_2(C_n)$ is contained in any intermediate polygon. This is due to the fact that E_n, E_1 do not appear in D defined in the proof. See Figure 25.

By the same method we can obtain the following results. Lemma 5 will be used in the proof of Theorem 2.

Lemma 4. *Assume furthermore that θ_3 is of type 1 and $S = \overline{V_3V_4} \cap \overrightarrow{V_1V_2}$. Then the conclusions in Lemma 2.6 hold with (b),(c) replaced by: (b$'$) throughout the transform V_1, V_2 and V_3 are fixed, and $E_3(C_n)$ is contained in any intermediate polygon; (c$'$) $\overline{V_2S} \cap C'_n = \phi$. (See Figure 29).*

Remark. A similar result holds if, instead of being the intersection point of $\overline{V_3V_4}$ and $\overrightarrow{V_1V_2}$, S is taken from $\overline{V_3V_4}$ and then kept fixed. See Figure 29.

Lemma 5. *Let $C_n \in \mathcal{P}_n$ with θ_2 of type 0 and θ_3 of type 1. Then C_n can be transformed to an equivalent $C'_n \in \mathcal{P}_n$ such that (a) $C'_n \cap \overline{V_2(C'_n)V_4(C'_n)} = \phi$, and (b) $< V_1(C'_n)V_2(C'_n)V_4(C'_n)$ is of type 0 and $< V_2(C'_n)V_4(C'_n)V_5(C'_n)$ is of the same type as θ_4.*

Proof: We claim that we may well assume (b) holds for C_n. Assume it is so temporarily. Then those portions of C_n that cross $\overline{V_2V_4}$ can be removed as before. This completes the proof.

The claim is verified as follows. By Lemma 3 we may assume $C_n \cap \overrightarrow{V_2S} = \phi$, where $S \in \overleftrightarrow{V_2V_1} \backslash \overrightarrow{V_2V_1}$. Thus $< V_1V_2V_4$ is of type 1. Since $(< V_2V_4V_5) > \theta_4$, we are done if θ_4 is of type 1. Otherwise, we obtain, by "cutting" C_n along $\overrightarrow{V_4V_3}$, $\tilde{C}$ with V'_2, V_4 lying below $\overrightarrow{V_5V_4}, \overrightarrow{V'_1V'_2}$ respectively. Note that V_4, V_5 are kept fixed throughout the transform. See Figure 30. This verifies the claim.

Q.E.D.

Now we start to prove Theorem 1.

Proof of Theorem 1. The proof is done by induction on n. The case $n = 3$ or 4 can be treated easily. This sets up the induction procedure. In the following we may assume $n \geq 5$.

If $\sum_1^n \phi_i = (n-2)\pi$ then there exist two consecutive angles, say ϕ_2 and ϕ_3, such that $\phi_2 + \phi_3 > \pi$. In the special case that C_n is convex, then ϕ_2 and ϕ_3 are of type 0. Thus E_2 can be moved arbitrarily close to P, where

$P = \overrightarrow{V_1V_2} \cap \overrightarrow{V_4V_3}$. See Figure 26. Without loss of generality we may assume

$$C_n = C_{n-1} \oplus \Delta_3(2) \tag{8}$$

where $C_{n-1} \in \mathcal{P}_{n-1}(\phi_1, \phi_2 + \phi_3 - \pi, \phi_4, \ldots, \phi_n)$ is the polygon formed by $\left(\bigcup_{4 \le k \le n} E_k\right) \cup \overline{V_1P} \cup \overline{PV_4}$, and $\Delta_3 \in \mathcal{P}_3(\phi_2 + \phi_3 - \pi, \pi - \phi_3, \pi - \phi_2)$ can be made arbitrarily small. Formula shows that $\mathcal{P}_n(\phi_1, \phi_2, \ldots, \phi_n) \neq \phi$ if $\mathcal{P}_{n-1}(\phi_1, \phi_2 + \phi_3 - \pi, \phi_4, \ldots, \phi_n) \neq \phi$. This proves (i). Since a similar formula holds for C'_{n-1}, (ii) follows from Lemma 1. The theorem is verified for c_n convex.

Some difficulty arises if C_n is not convex. Even if ϕ_2 and ϕ_3 are of type 0, there is not guarantee that $E_2(C_n)$ can be moved as before. Certain operation is needed to "evacuate" C_n from ΔV_2V_3P. This can be achieved by use of Lemmas 1-4. From now on we assume C_n is not convex.

Part (ii). Since $\sum_1^n \phi_i = (n-2)\pi < n\pi$, there exist two consecutive angles, say ϕ_2 and ϕ_3, such that ϕ_2 is of type 0 and ϕ_3 of type 1.

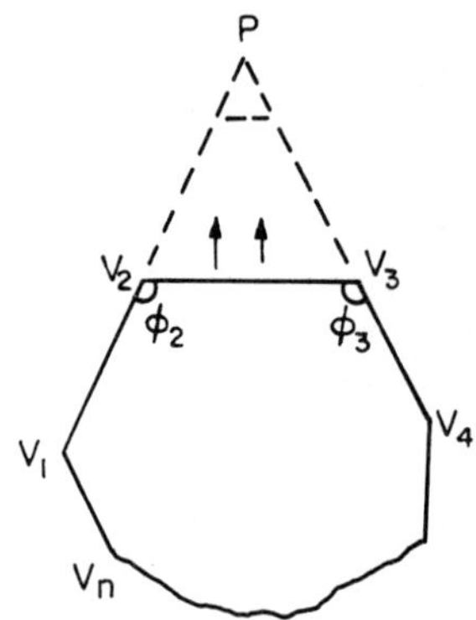

FIGURE 26. Edge E_2 can be moved arbitrarily close to P.

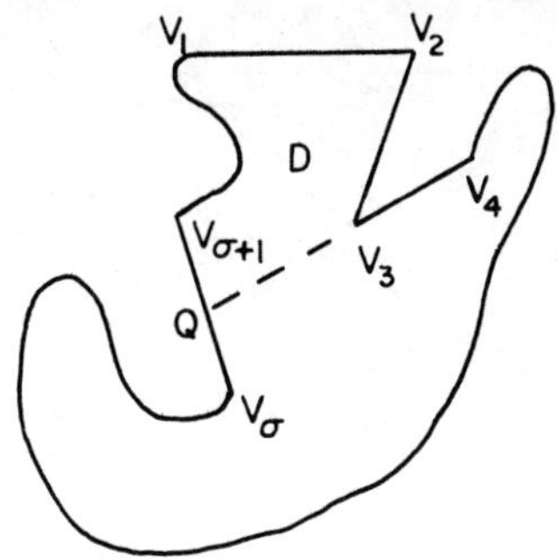

FIGURE 27. D can be shrunk arbitrarily close to Q.

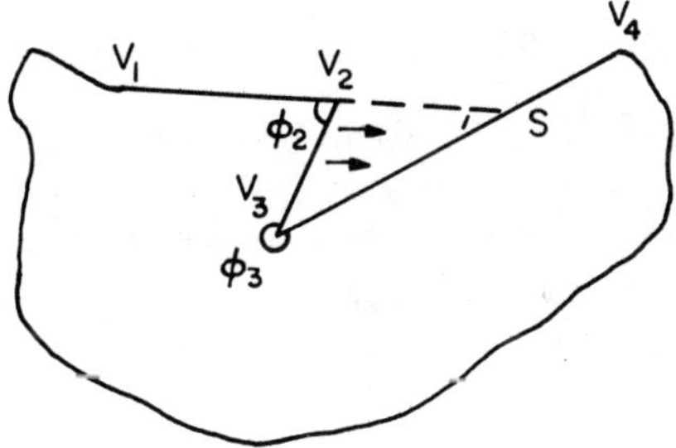

FIGURE 28. Edge E_2 can be moved arbitrarily close to S.

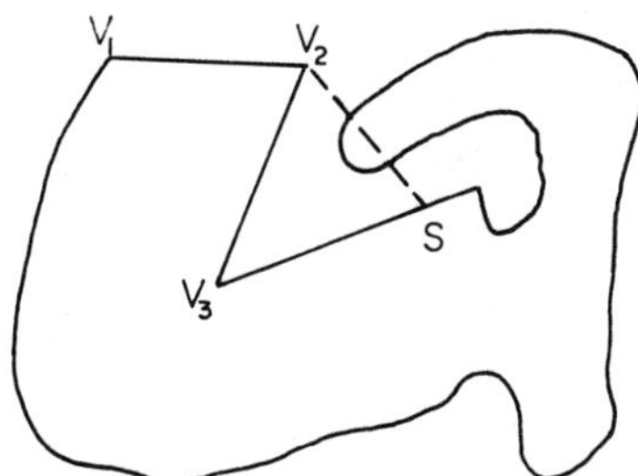

FIGURE 29. S is fixed once chosen. Those portions of C_n that cross $\overline{V_2S}$ can then be removed while keeping V_2 fixed.

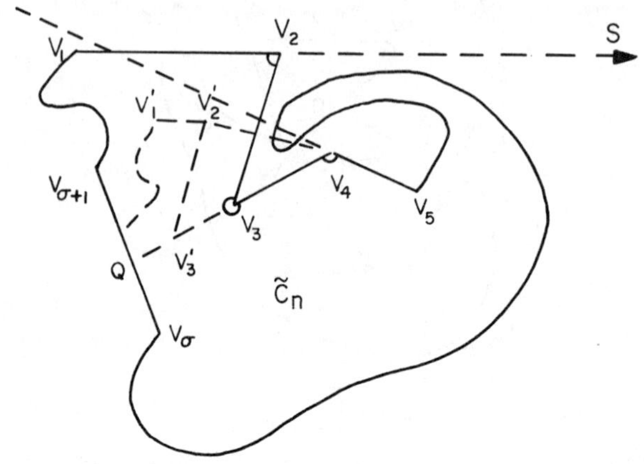

FIGURE 30

Case 1. $\phi_2 + \phi_3 > 2\pi$. By Lemma 2 the polygon D formed by $\left(\bigcup_{\sigma<k<2(mod n)} E_k\right) \cup \overline{V_3Q} \cup \overline{QV_{\sigma+1}}$ can be shrunk arbitrarily to Q (see Figure 27). Without loss of generality we may assume $\overline{V_1V_2}$ and $\overline{V_3V_4}$ intersect at a point S. By Lemma 4 we may assume further that $C_n \cup \overline{V_2S} = \phi$. Since E_2 can now be moved arbitrarily close to S, we have

$$C_n \sim C_{n-1}\theta\Delta(2), \tag{9}$$

where $C_{n-1} \in \mathcal{P}_{n-1}(\phi_1, \phi_2 + \phi_3 - \pi, \phi_4, \ldots, \phi_n)$ and $\Delta_3 \in \mathcal{P}_3(\phi_2 + \phi - 2\pi, \pi - \phi_2, 2\pi - \phi_3)$ can be made arbitrarily small. See Figure 31. Since a similar formula $C'_n \sim C'_{n-1}\theta\Delta_3(2)$ holds and $C_{n-1} \sim C'_{n-1}$ by the induction hypothesis, $C_n \sim C'_n$ by Lemma 1.

Case 2. $\phi_2 + \phi_3 \leq \pi$. For the convenience of discussion we take V_2 as the origin and $\overrightarrow{V_2S}$ the positive x axis, where $s \in (\overleftrightarrow{V_2V_1} \backslash \overrightarrow{V_2V_1})$. Let (x_iy_i) be the coordinate of V_i. By "cutting" C_n along $\overrightarrow{V_4V_3}$ we may assume without loss of generality that $x > 0$. See Figure 31. By Lemma 3 we may assume further that $C_n \cap \overline{V_2S} = \phi$. Let $x_\tau = \max 1 \leq k \leq nx_k$ and has the smallest subscript. See Figure 31. Then $\tau \neq 2$ and ϕ_τ is of type 0. Define $P = \overrightarrow{V_2S} \cap \overrightarrow{V_{\tau+1}V_\tau}$. Consider the polygon D formed by $\left(\bigcup_{2\leq k\leq\tau} E_k\right) \cup \overline{V_\tau P} \cup \overline{PV_2}$. Since $< V_3V_2P, < V_2PV_\tau$ and $< PV_\tau V_{\tau-1}$ are all of type 0, it follows from Lemma 2.5 that D can be shrunk arbitrarily close to P within $N(D; \delta; \overline{V_2P}, \overline{PV_\tau})$ and with P fixed throughout the transform. Therefore, edges $E_{\tau+1}, E_{\tau+2}, \ldots, E_n$ of C_n are not affected without loss of generality that $(\bigcup \tau \leq k \leq nE_k) \cap \square V_1V_2V_3T = \phi$, where $\overleftrightarrow{TV_3} \parallel \overleftrightarrow{V_1V_2}$ and

$T \in \overline{V_1V_n}$ or $\overleftrightarrow{V_1V_n}$ depending on whether ϕ_1 is of type 0 or not. See Figure 32. Note that a slight perturbation of E_n might be necessary for making $\left(\bigcup_{\tau \leq k < n} E_k\right) \cap \overline{V_1T} = \phi$. It does not necessarily follow that

$$\left(\bigcup_{4 \leq k < \tau} E_k\right) \cap \Box V_1V_2V_3T = \phi. \tag{10}$$

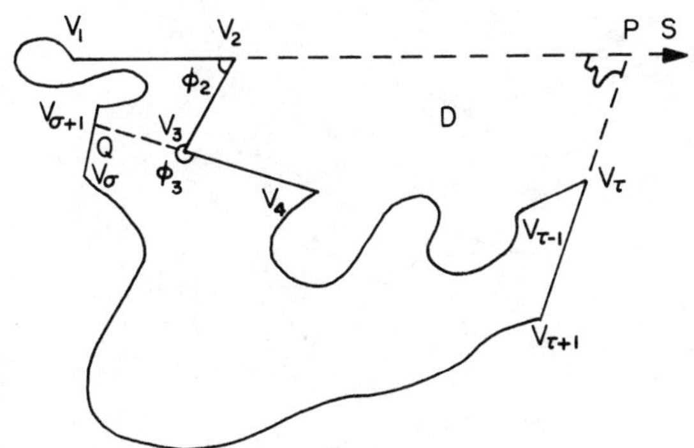

FIGURE 31. $x_4 > 0$ and D can be shrunk arbitrarily close to P.

See Figure 32. However, by Lemma 3 we may assume $D \cap \overline{V_3T} = \phi$. See the remark after Lemma 3. In particular, (10) holds. Because of Lemma 3(a), $\bigcup_{\tau \leq k < n} E_k$ are kept away from $\Box V_1V_2V_3T$ in obtaining (10). We finally may assume $(C_n \backslash E_n) \cap (\overline{V_1T} \cup \overline{TV_3}) = \phi$. Then edges E_1 and E_3 can be moved arbitrarily close to T. See Figure 32. This implies

$$C_n \sim \begin{cases} C_{n-1} \oplus \Delta_4(1), & if\ \phi_1 \text{ is of type } 0, \\ C_{n-1} \oplus \Delta_4(1), & if\ \phi_1 \text{ is of type } 1, \end{cases} \tag{11}$$

where $C_{n-1} \in \mathcal{P}_{n-1}(\phi_1, \phi_2 + \phi_3 - \pi, \phi_4, \ldots, \phi_n)$, and Δ_4 is an arbitrarily small 4-gon with $V_1(\Delta_4) = T$.

Since a similar formula holds for C'_n, the conclusion $C_m \sim C'_n$ follows from (11) and the induction hypothesis.

Part (i). This follows easily from (9) and (11) by using the induction.

This completes the proof of the theorem.

Q.E.D.

Theorems 2 and 3 can be proved easily now. Assume $C_n \in \mathcal{P}_n(A)$, where $A = (a_1, a_2, \ldots, a_n)$ is a $0-1$ n-tuple. Let $1 \leq i_1 < i_2 < \cdots < i_m \leq n$ be

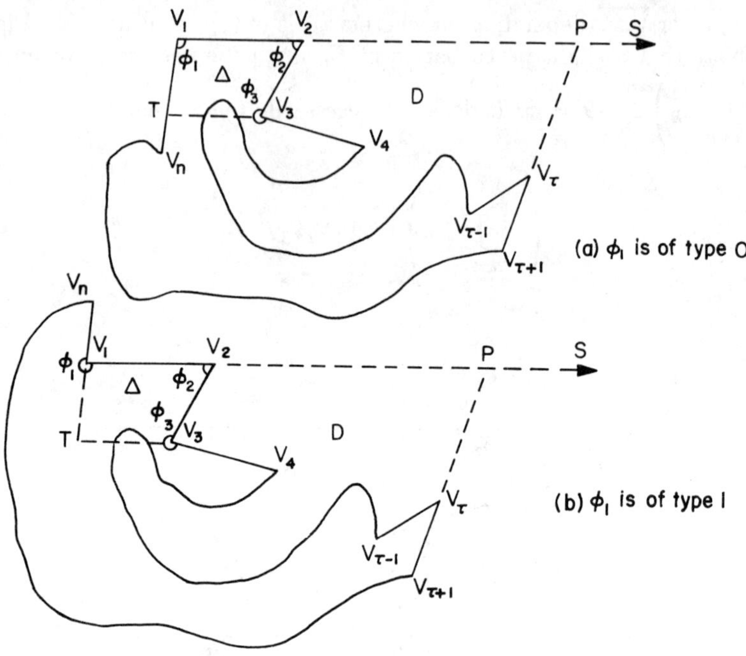

FIGURE 32

those index i such that $a_i = 0$. For convenience we take $i_1 = 1$. Note that $m \geq 3$, because $\sum_1^n \phi_i = (n-2)\pi$ implies C_n has at least three angles of type 0. Let D_m be a regular m-gon. For each $1 \leq j \leq m$ cut away along edge $E_j(D_m)$ a convex $(i_{j+1} - i_j + 1)$-gon B_j. (This happens only when $i_{j+1} - i_j > 1$.) By requiring B_j to be very thin, i.e., close to $E_j(D_m)$, we obtain an n-gon C_n^* in $\mathcal{P}_n(A)$. See Figure 33 for an example. It is clear that Theorem 2 has the following equivalent form:

Theorem 2.2'. $C_n \overset{s}{\sim} C_n^*$.

Proof: The proof is done by induction on n.

Since the theorem is obviously true for C_n convex, it suffices to consider the nonconvex case. Then there exist two consecutive angles, say θ_i and θ_{i+1}, such that θ_i is of type 0 and θ_{i+1} of type 1. By Lemma 5 we may well assume $C_n \cap \overline{V_i V_{i+2}} = \phi, < V_{i-1} V_i V_{i+2}$ is of type 0, and $< V_i V_{i+2} V_{i+3}$ is of the same type as θ_{i+2}. Let $C_{n-1} \in \mathcal{P}_{n-1}((a_1, a_2, \ldots, a_i, a_{i+2}, \ldots, a_n))$ be the polygon resulted from replacing edges E_i, E_{i+1} of C_n by $\overline{V_i V_{i+2}}$. Since v_{i+1} can be made arbitrarily close to $\overline{V_i V_{i+2}}, C_n$ is s-equivalent to C_{n-1} minus a very thin triangle Δ_3 sitting on $\overline{V_i V_{i+2}}$. See Figure 31. Express

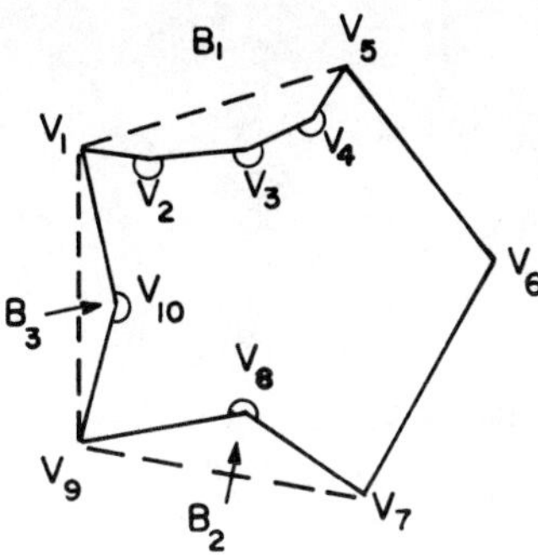

FIGURE 33. A representative element of $\mathcal{P}_{10}(0,1,1,1,0,0,0,1,0,1)$.

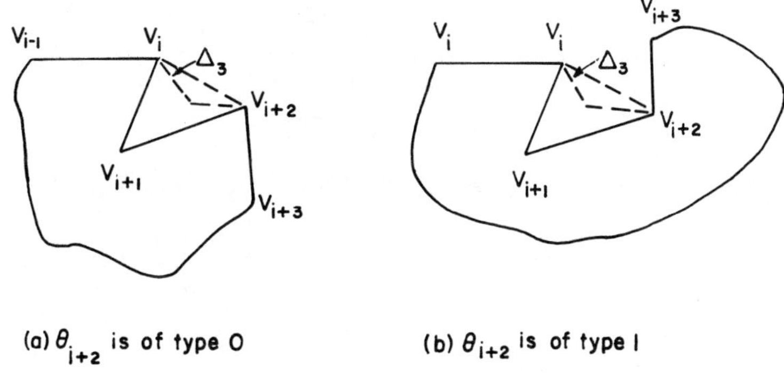

FIGURE 34. Δ_3 can be made arbitrarily close to $\overline{V_iV_{i+2}}$

this roughly by $C_n \overset{s}{\sim} C_{n-1} \ominus \Delta_3[i]$. The conclusion then follows from the induction hypothesis (applied to C_{n-1}) and the following s-equivalence version of Lemma 1: if $C_k \overset{s}{\sim} C'_k$ and Δ_3 is sufficiently thin, then $C_k \ominus \Delta_3[i] \overset{s}{\sim} C'_k \ominus \Delta_3[i]$. The proof of this version is similar to that of Lemma 1 and is omitted here.

Q.E.D.

Proof of Theorem 3. It suffices to consider the case $C_n \in \mathcal{P}_n$. By Lemma 2.2 we may assume $C_n = C_n^* = D_m \ominus (\sum_1^m B_j)$. Now push each B_j outward to obtain a convex n-gon. See Figure 35 for B_j to be a triangle. Since any two convex n-gons are w-equivalent, the theorem is proved.

Q.E.D.

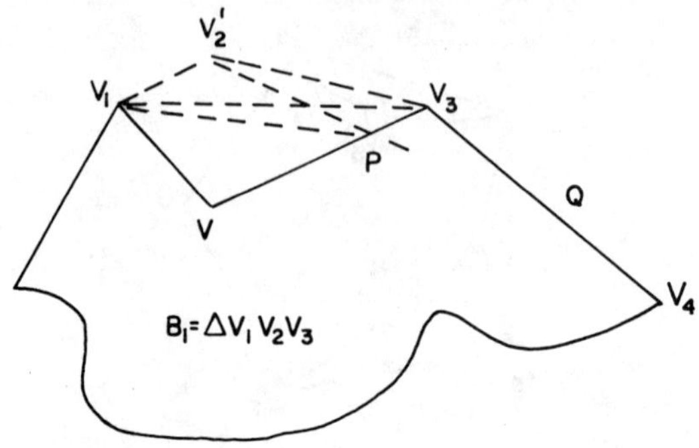

FIGURE 35. V_2 is moved to V_2' in several steps:

1. $\overline{V_1V_2} \to \overline{V_1P}$
2. $\overline{PV_3} \to \overline{PQ}$
3. $\overline{V_1P} \to \overline{V_1V_2'}$
4. $V_2'Q \to V_2'V_3$